Arnaldo Arduini

# Principles of Theoretical Neurophysiology

Springer-Verlag
Berlin Heidelberg New York
London Paris Tokyo

Professor Dr. ARNALDO ARDUINI
Scuola Superiore di Studi Universitari
e di Perfezionamento
Via Carducci 40
I-56100 Pisa

Library of Congress Cataloging-in-Publication Data. Arduini, Arnaldo. Principles of theoretical neurophysiology. Bibliography: p. 1. Neurophysiology – Philosophy. 2. Biophysics – Philosophy. I. Title. [DNLM: 1. Biophysics. 2. Brain–physiology. WL 300 A677p]
QP356.A77 1987 612'.82'01 87 – 4873

ISBN-13: 978-3-642-71470-2 e-ISBN-13: 978-3-642-71468-9
DOI: 10.1007/978-3-642-71468-9



Softcover reprint of the hardcover 1st edition 1987



Media conversion: Daten- und Lichtsatz-Service, 8700 Würzburg

2125/3130-543210

*To Mariagrazia,*
*my loving and encouraging wife*

# Preface

The present book has two origins, one very remote, the other nearer and more contingent. The first goes back to the time when I initiated my career as a neurophysiologist in Pisa, a small town with the advantage of a highly stimulating atmosphere created by two famous institutions, the University and the Scuola Normale Superiore. It came quite naturally, then, while engaged in experimental work, to start brooding over the possible analogies between neurophysiological problems and those of the physical world. This slowly induced me to become less interested in the solution of the innumerable specific problems presented by the brain, and more in the general principles on which the brain function might be based. Certainly, for several years I had no clear idea of my purposes, or of the difficulties I could encounter in the task. However, it was clear enough that there ought to be a first, indispensable step: the search for methods of quantification of nervous activity, the sole way of allowing predictions about its behavior. At first, I somehow followed the fashion of the time, experiencing the impact of information theory on neurophysiology, but soon this was revealed as unsatisfactory, since it was only one aspect of the problem, and what I was interested in was not a way of describing the flow of information, but rather, the laws of the machinery.

Living bodies are made of parts organized in a "whole", all of them, from unicellular algae to the hugest of beasts. The characteristic of all living beings is that they are not simply the sum of billions and billions of molecules or of cells: the "whole" counts for more than the simple sum of its constituents, because of an organization which enhances the contribution of each part and causes the emergence of properties not possessed by the individuals. This aspect has not discouraged most researchers, even if the complexity of the brain's organization seems to reach unexpected degrees. As a consequence, the neurosciences have developed the procedures through which, piecemeal and painstakingly, the fine structure of brain elements and their interconnections are being studied. For my part, I never believed the organization of the brain to be a problem of structure alone. I believe there is a general plan to be represented in some frame of thought, not requiring detailed knowledge of the individual elementary connections, but based on the general properties of a class of complex systems to which our brain belongs. This brought me nearer to physics and mathematics than I had foreseen, but I have been rewarded discovering the innumerable affinities, and also the true identities of the general properties of the system "brain" with those of other systems in the inanimate world.

Yet, recently, I had to face the difficulty in stating my problems and my standpoint in a convincing way for a physicist, because the somewhat raw matter I had in mind still needed systematic revision and an order. While engrossed in the latter job, I suddenly became aware that what I had been missing was actually there, the outline of a general theory of neurophysiology, a set of principles of the brain function, and this was how the present book took shape. As it is, it can only represent a first contribution to a comprehensive theory of neurophysiology: perhaps, part of it has no better claim as an approach than that of a "verstaendlische" approach, a set of hypotheses helping to understand the facts, but certainly with respect to another part it has been able to make predictions, and so far the outcomes of observation support the theory.

The purpose of this book is to make an understandable story for all concerned, neuroscientists and physicists, but, since I have tried to explain how brain and physics may meet from uncommon viewpoints and with as little specialized jargon as possible, I am afraid everyone, neuroscientists and physicists, will be discontented, and will have to bear with the writer throughout the book.

There is also a utilitarian finality in my endeavours. The fields of the neurosciences and of the exact sciences, such as physics and mathematics, have so far had only occasional and superficial points of contact. The reasons are clear: the diffuse belief that the respective fields are too far apart in every respect, that the conditions of living systems are not reproducible with sufficient predictability and precision to submit them to exact treatment, that detailed knowledge of the forbiddingly complex structure of the brain is a necessary prerequisite. I think I can demonstrate that all these are prejudices. My hope is that I shall eventually succeed in attracting the interest of physicists and mathematicians towards an approach to neurosciences in which, if not all, at least the main biological hindrances to an exact treatment are sidestepped. Then, the neurosciences will gain the full advantage of the specific experience of the exact sciences, and the latter will find in the brain an unexpectedly fruitful field for testing their theories. As a corollary to these purposes, I am proposing, as well, what I believe might become a general, or common approach to different types of neuroscience investigation, using language which is simple and very common in physics and mathematics.

For the same reasons, I have also used the matter in a seminar course, as part of the teaching program of neurophysiology for both medical and non-medical students. The course was found to meet their need for a generally comprehensive, synthetical view of the mechanisms of the basic cerebral processes. The matter is ordered in the book as it was in the seminars, the aim being common to both.

Pisa, Spring 1987 ARNALDO ARDUINI

# Contents

# Part I General Properties of the Brain

# 1. Introducing the Problem

*Faust:*

Habe nun, ach! Philosophie,
Juristerei und Medizin,
Und leider auch Theologie
Durchaus studiert, mit heissem Bemühn.
Da steh'ich nun, ich armer Tor!
Und bin so klug als wie zuvor;...
(*Goethe*, *Faust*, v. 354–359)

## 1.1 Standpoints

Neurological research has expanded and diversified to include neurophysiology, neurochemistry, neuropharmacology, psychology, neuropsychology (an offspring of physiology and psychology) and, of course, neuroanatomy, pathology, and clinical neurology. Each of these branches has developed knowledge in bulk, with its own language and specialized journals. A huge amount of data has been produced at an exponential rate in the last three decades, demonstrating that the field is lively and bursting with activity. These data also show that the brain is an enormously complicated object and, therefore, that more is necessary than separated detailed analyses: differentiated, yes, but simultaneous and coordinated approaches.

Dedicated scientists have been rewarded with substantial achievements, aided by increasingly improved technical support, but mainly thanks to their individual ingenuity. This has created what is called a boom: with the mail we are no longer receiving the monthly accumulated data, but also the weekly ones. Anyone wanting to keep himself up to date in more than one restricted subject feels dejected: the languages are different, the information is differently ordered, clinical and experimental data do not fit together; neurochemists and pharmacologists are working independently from the others, discovering new molecules every other day. Moreover, technical jargon is pervasive and the time is not sufficient for accurate and critical studies, a situation calling for some change.

Certainly, much progress is being made in the whole field, but it is very hard to perceive in it a "direction" that is not simply the expansion of one of the subfields. If a general tendency can be perceived, it is in the uniform attitude with which the neurophysiologists are approaching the neurosciences as the archeologists do buried civilizations: they dig and slowly uncover a gigantic plan of structures, a myriad of channels connecting centers, and unveil the traces of unknown activities. Little or no attention is paid to the fact that beyond what is more or less immediately perceived by our senses as scientists there is another reality, that of the principles which form the basis of our science, which by that very fact are deeply rooted in the human mind. This book endeavours to unveil these principles. Reason itself suggests that another reality also exists, independent of the human mind, a reality of which the mind can perceive only a shadow. However, in the course of the following chapters we shall not be concerned with the latter reality: only in the Appendix may one find hints of future discussions of it.

If we compare the development of the neurosciences with that enjoyed by other scientific domains, such as mathematics, physics or chemistry, we are struck by the beautiful simplicity of the main lines of thought, which can, generally, be followed beyond the detailed analysis, the technicalities, and the jargon – all the tools one has to employ in research. In the neurological sciences we presently have no main line of thought to follow, thus we are left with the technicalities, the search for extreme details, and the jargon which renders most of the scientific papers incomprehensible to non-adepts. In other words, neuroscience lacks a unifying philosophy of research. I am deeply convinced that if science wants to really progress, hypotheses need to be made and quickly tested for their veracity, yes, but the hypotheses must stem from an all-embracing frame of thought, from the very principles of the discipline. They cannot be simply tactical, "working hypotheses".

I like to recall in this connection Einstein's words: "Eine Theorie kann an der Erfahrung geprüft werden, aber es gibt keinen Weg von der Erfahrung zur Aufstellung einer Theorie", that is, a theory can be tested by experience, but there is no way leading from experience to theory. These words clearly state the roles of tools, methods, experiments, and theory, attributing to the latter the primacy, the guidance of the task of assessing the truth in the natural world. Thus, the viewpoint of several scientists looks untenable: Karl Jaspers, for instance, in his "Allgemeine Psychopathologie" minimizes the necessity of a general philosophy and of a general theory, and denies even the possibility of constructing one in his field. We must, however, be convinced that everything occurs by virtue of some principle, otherwise we could picture ourselves in a perfectly chaotic world. Perhaps, also under the influence of Jaspers's overwhelming personality, no conclusive trials have been made, not only in psychopathology, but also in other neuroscientific fields, such as basic neurophysiology. One has to concede that the more detailed the analysis, and the more data is accumulating, the harder it is to preserve a general view of the problems. Presently, overall knowledge of the brain could result only from ordering a heap of piecemeal data, painstakingly gathered by a large number of investigators, who individually do not have the time to obtain sufficient insight into other neurological branches. This, of course, besides hampering an overall view of neural problems, also prevents one from understanding their relationships to other problems. In more than 35 years as a neurophysiologist, I have gained the profound conviction that the only realistic approach to the multitude of problems presented by the brain's functions can only be a broadly naturalistic one. One should not refrain from approaching our problems with the physicist's, the chemist's, or the mathematician's eyes. We should never forget that we have to consider the living being, and all the manifestations of life, as part of Nature, of which they share all the laws. This first assumption is the basis for the following chapters.

I am aware that there might not be consensus when this assumption is boldly taken as a starting point for practice. Notwithstanding the efforts on different sides, biological problems, those of the brain in particular, are always considered unapproachable. While there is general agreement, I think, on the fact that all the events of life should be described in physical and mathematical terms, there is, certainly, a good deal of skepticism about the actual possibility of reaching the goal. In my opinion, it might be primarily a problem of approach, for want of that general frame of principles within which the physico-mathematical description should find its place. This is certainly true, if not for all the fields of biology, at least for the neurosciences, for the brain. Everyone is ready to concede that the complexity of biological systems has, generally, no equal among inanimate ones, and thus that absolute certainty and precision of description are not to be expected, even on the macroscopical level. Accepting the point of view of a famous physicist (B. D'Espagnat), we should not always require absolute certainty in science. A quasi-certainty could do, at least to start with, provided we end up with a widely embracing and self-consistent vision. This is exactly what I hope to be able to produce – a picture of our brain's ways of functioning that surely can be further improved, but which represents at least a sufficiently reasonable basis to begin with. The book has been written at an early stage of development of the ideas with this last consideration in mind. It is intended primarily as a stimulant (first for the writer himself) to pursue a line of investigation, to look for confirmation of experience, that can only result in improving the focus of what has been simply outlined here and presented as a still hazy picture. There is a considerable amount of work ahead to be done, both theoretical and practical, demanding that not this single man, but many different teams become involved in the task.

If we want attain that overall perspective, to construct a logical framework for all the problems of our brain, then we must accept some changes in our general attitude towards those branches of science not directly dealing with neurology. It must be clear that "finding a frame for the brain" is not intended simply as a description of its states, but as the definition of the principles determining the chain of causes and effects which constitute its being, in other words the laws of its functioning.

My assumption entails that the use we generally make of the body of knowledge pertaining to other branches of science must not be limited to the superficial exchange of information about the more recent trends in research. Let's abandon superficial curiosity. The vast majority of us tend to consider the other disciplines as at best a source of new tools, methods, and technical improvement. In other words, we generally ask ourselves how they can be of any use in "our" research, as if our problems were of a different nature from theirs. Instead of trying to simply exploit material achievements, our own thinking needs to be re-addressed when we are confronted with other branches of knowledge, because the many difficult problems presented by brain function can be (but I prefer to say that they must be) approached and solved as particular cases within classes of similar problems in the other realms of nature. The notion that living systems are vastly different from other systems in nature is still lingering in the back of our minds, helping to limit our excursions in other fields to only the search for useful "models", either mathematical models for computing some statistics, or physical models for reproducing in man-made objects some of the properties of nerve cell circuits. The undeniable complexity of the latter has led us astray, first by convincing us that they cannot be attacked at all as they are, secondly by luring our efforts into an endless search for methods of simplification, in the hope that somewhere and at some time we shall strike the right model. Thus, we behave as if there were no objects in the physical world with a complexity comparable to that of our brain. Perhaps this is so, nonetheless there are fairly complex systems among man-made devices, for instance the most sophisticated computers. These are enormously complex, made up of innumerable parts and functions, but there are laws governing their functioning, and they can be described through these laws.

The point is that there are no laws which are specific to one complex object and not valid for another one. The higher number of functions and of degrees of freedom allowed in a system may impose an additional burden of computation, but, in principle, should not require different laws, unless different properties emerge. If the laws are of a sufficiently general character, as every law should strive to be, they will characterize the brain as a particular case, perhaps the limit one, within a class composed of comparably complex systems. Having outlined my starting point, I will look into other fields, not with the purpose of finding simpler models for our brain, but rather, the reverse. I will try to determine which of the other fields "contains" the brain as one of its objects, as a living model of its laws.

There are also practical considerations which have lately begun urging me in the direction outlined. As soon as the technical progress of the last decades allowed extensive investigations of the brain's activity, the neurosciences felt the pressure, more than other medical branches, towards a quick exploitation of laboratory data in order to create and improve diagnostic and therapeutic methods. The pressure has increased enormously in the last few years since neurochemists and pharmacologists have extensively investigated the molecules involved in the different aspects of brain function. Fostered by big concerns, this chemico-pharmacological line of investigation, while representing the newest, richest and most promising approach, unveiling unexpected aspects of nerve function, introduces at the same time further differentiation within an already strongly differentiated field. Unfortunately, there has so far been no comparably strong attempt at connecting the newer with the older lines of research, a fact that should make everybody aware of a rapidly widening gap, and makes me feel that the time is ripe for a thrust into the problem of finding a unified scheme of neural function.

## 1.2 Procedures and General Plan

*Mephistopheles:*
Nicht Kunst und Wissenschaft allein,
Geduld soll bei dem Werke sein.
(*Goethe*, *Faust*, v. 2370–2371)

In that unified scheme, all the results from any line of research should find a place, provided their meaning is translated into a unified language (units, code, or representation). This should permit an immediate appraisal of the bearing of every type of results upon the general picture of the brain's function. Moreover, if that scheme is based on the general laws governing similar complex systems, the results become largely predictable. If we look into it more closely, however, the latter condition is equivalent to an outline of a general theory of brain function, not simply a unification of methods and language. It would seem logical to start with the (apparently) simpler task of trying to define the code through which the brain can tell us the details of its operations in relation to its anatomical structure, its chemical composition, and the information it is handling. However, the procedure cannot be that straightforward. The principle of unification is based on the concept of "transformation", the concept implying that the physiological meanings and the physical measures (the data are always actual measures of some parameters of brain function) are so defined that the same meanings and measures are preserved after the transformation. Depending upon the frames we shall consider, there are different possible transformations. As we proceed to specify the complex laws of brain function, the representation itself will be better specified, so we shall have to wait until the very end to know in what form the theory shall be represented.

In this procedure we encounter an obstacle. Since we set out to build a theory, that is to discover the general "philosophy" of the brain, we come upon a long-debated (and almost forgotten) question, i.e. how far the brain can investigate and, thus, gain objective knowledge of itself. Let this question not stop us. It has been raised with reference to the innermost, and still rather mysterious ways in which our brain constructs its thinking, the ways it constructs abstractions and concepts not belonging to the physical world and of which the physical world cannot possibly give a picture, a representation, or a hint. We are not (at least not yet) aiming at such perfect knowledge of our brain. We shall be satisfied with establishing the general principles of its mechanisms, a kind of knowledge far less detailed and complete than the knowledge of specific mechanisms (such as motility and sensation) but essential to pursue their investigation. There is only one way to reach this. A man cannot see his own face, but has to look at it in a mirror and compare it with those of others. The same, in a sense, is true for the brain. To gain primary knowledge of itself, to understand its own ways and laws, the brain has to mirror itself in the ways and laws of nature outside. So, in practice, we shall follow a purely empirical standpoint, assuming that inasmuch as the brain can be described like any other of the complex systems found in nature, it must also be subject to the same laws and fate. (In Sect. 1.3 we will consider an alternative point of view).

Assuming this starting point is justified, we have no choice: we have to deal with a physical system made of innumerable parts, having properties characterizing it at a micro- as well as at a macroscopic level, with reciprocally influenced components comparable to long-range and short-range forces, therefore we must ascertain to what extent the laws of nature, of physics first of all, condition the function of the nervous system. Why physics, out of all, is easily understandable. All natural phenomena can be essentially described in physical terms, the other sciences having developed later. While we can trace the logical path leading from the physical to the chemical, the thermodynamical and the biological descriptions of nature, there is no logical procedure retracing the path.

With its many branches, physics deals with systems made of a large number of components, either moving or not, and has developed corresponding laws to describe them. What gives primacy to physics, besides having originated the other sciences of nature, is its being extensively entwined with mathematics, from which it derives the rigor of its logic. We will first attempt to identify the physical properties of our brain. Probably, my fellow neurophysiologists will be very surprised and annoyed by the absolute absence of any anatomical description. The latter would certainly be altogether superfluous here, since what we are interested in are the basic structural aspects, i.e. those through which we can compare the brain with other (non living) physical systems as members of the same class. It is obvious that in this perspective the problem of the detailed structure of the brain is not relevant. This should not convey the impression that our talk will be concerned with an abstract brain or a simplified model. What I hope to be able to demonstrate is that our real brain has fundamental properties not based on any particular anatomical structure. To avoid even the suspicion that our object is not the real brain, I will refrain from what would seem most natural to physicists and to most neurophysiologists, i.e. I will not compare the brain with automata of any complexity, although the latter have become our companions in research because of the similarities of function, if not of operation. Thus, our attention will always be directed to fundamental principles and general laws, not to their practical applications, which would represent only models which we are not looking for. For this reason, the material of the book is not ordered as one would a systematic exposition of neurophysiology, but follows another logical sequence, dictated by the principles themselves.

The book has been divided into two parts: the first one is to give non-biologists some of the essentials of the brain's function, indepedent of anatomical descriptions and detailed physiological knowledge, while the second part should direct both neuroscientists and physicists towards areas where their respective disciplines may meet, with problems and language in common.

The binome structure function constitutes the main subject of a lengthy discussion dedicated to "organization", the principal issue of the book. Organization is strictly bound to structure, which it needs, but with which it cannot be identified. Again, organization is intended primarily in functional terms, even if I cannot and do not want to dismiss the presence of the extremely rich maze of connections binding the innumerable parts of which the brain is made.

The chapters are ordered in such a way to progressively lead the reader to get acquainted with the relevant functional characteristics of the brain as a physical system. I have tried to avoid the term "analogy" since my aim is to demonstrate that brain functions can be described through exactly the same principles and laws we find in statistical and quantum dynamics, in classical mechanics, in thermodynamics, in the mechanics of continua, in relativity. In many cases, however, an analogy is all we can obtain. In order to be able to do all this, we shall also have to dwell on the nature of the variables used to describe the brain, with critical appraisal of the methods and the current interpretation of data. Obviously, the selection of physiological variables is a critical point, the difficulties residing in the poor definition of the variables used almost universally to represent the different parameters of nerve cell activity. The problem is not exhausted in a single chapter, being resumed whenever the previous definitions become inadequate and need to be made more precise as the discussion progresses, (as with the type of activity equated to motion, the neural forces, and the "states" of the brain).

As is almost traditional in neurophysiology, statistical reasoning has a large consideration in this book, and I have tried as much as I could to point out the limits of the statistical approach to brain problems. I am willing to concede that there is some bias on my side, since a preference for more deterministic representations to be found in the following chapters is certainly based on my previous failure with the usual statistical ones. However, the resulting overall view of the brain is not mechanical in the pure classical sense, rather it follows quantum mechanical principles, which, perhaps too boldly and not without hesitation, I have transferred to the nervous system. Surely this may look like a logical fault, but it has something to do with the expedient change of perspective I had to implement in my neurophysiological approach. The extremely large number of degrees of freedom of our nervous system does not allow a precise definition of all the variables involved, so we are forced to think of their states as being condensed, so to speak, in a restricted number of their functions, into some kind of "state functions", which thus acquire the significance, if perhaps not the exact form, of the state functions in quantum mechanics.

The final picture corresponding to the theory I am proposing for the nervous system is completed with ideas taken from all the chapters, including the unusual (for the physiologists) Gaussian representation, the theory of continua and some relativistic hints. At the end, however, the reader may get the impression that there has been "much ado about nothing", that what one has obtained practically from all the talk about physical principles of brain function is still insignificant. Certainly, a consensus should follow when the few principles outlined here are put to test in the practice of everyday research, and I hope it will. Since a very large number of variables are required for a suitable description of the brain's functions, a large part of the hindrance will be presented by the technical difficulties of a practical application. Of this I am thoroughly aware, having personally experienced it, though working with only few variables at a time, but I am deeply convinced that everything can be done with some ingenuity and a great deal of patience.

## 1.3 The Beyond-the-Physics Problem

*Mephistopheles:*
Wer will was lebendigs erkennen und beschreiben
Sucht erst den Geist herauszutreiben,
Dann hat er die Teile in seiner Hand,
Fehlt leider! nur das geistige Band.
(*Goethe, Faust*, v. 1936–1939)

At the end of the book I have added an Appendix, which does not contain the usual detailed treatment of topics, as mathematics, of the preceding chapters, but represents the natural consequence of the initial standpoint, the need for a philosophy of the neurosciences fitting the frame of the laws governing our Universe in general.

While a description of the brain's functions in purely physical, mechanistic terms may be found satisfactory by a good many students and even by laymen, certainly someone will feel dissatisfied and ask whether we are sure that a description of our brain can be exhausted in terms of physical relations, whether in our brain there is not anything "more" and, if so, why not talk about it. Whether what the neuroscientist investigates with his probes is actually all there is to know, or is only a part of a superior "unity" together with some non-material "entity", has always been, and will long remain, a scientifically unanswered question.

How the neurophysiological aspects can merge with the metaphysical ones has recently been extensively treated in a book by a philosopher and a neurophysiologist, Sir Karl Popper and Sir John Eccles. Unfortunately, our present extremely imperfect knowledge of the brain does not let us see to what philosophical depths investigations will lead at a future, maybe not too distant, time. Max Born, the Nobel Prize physicist, maintained that theoretical physics is actual philosophy (a view, I think, everybody can share), a standpoint reminiscent of the natural philosophers of antiquity. Whether there will be, after this epoch of ours, a new development in philosophy, with another Aristotle, Thoma of Aquino, Descartes, Leibnitz or Kant, no one can foresee. To what extent the penetrating and abstracting power of the human mind will be able to unveil its own depths, and its science the mysterious depths of the Universe around, no one can foresee either. However, when certain borders are approached, science by itself does not feel entitled to say one word more. The physicists have been able to retrace the different stages of the formation of this Universe down to the first .0001 seconds of its life, but they do not dare to venture much more about the state of matter at that moment and before (if there was a "before") or about what relations were there between space and time (if one may still talk of space and matter and time) when the Big Bang occurred. One reaches the point where words fail, because one does not know exactly what the concepts are, so physicists and philosophers are still debating the roots of reality and of its independence from the human mind which conceives it, the nature of its "being", the "esse". The problem is the same for our brain, except that, unfortunately, we have not reached as far as physicists have done with the Universe's story. But the "Big Bang" was a unique event, about ten thousand million years behind us, whereas our brain is still here and now, during the short span of our life, one for every individual, and whatever is there besides physical relationships must still be and remain there up to the end, within our reach. Would it be too daring to assume that its "traces" should be detectable, inasmuch it has to "use" the physical brain in some way? Perhaps, if our theories were based on laws general enough, we could find out where and how that "something more" (through its interfering, interacting, and co-operating with the nervous tissue and its organization) might turn out to be an essential of the normal brain's life, making with it a "unity". More than for all the problems tackled in the rest of the book, the reasoning and the approach followed in the Appendix must include the brain and its relationships with its own body, the outside world, and the whole Universe as well. While this is also a trend, a sign of our time, it might turn out to be a true necessity if we want to attain complete understanding of our brain as a living system, as the essence of Man. Thus, the significance of the Appendix will be more metaphysical than physico-neurological, not the advancement of hypotheses, since they certainly cannot yet be submitted to experimental test, but simply the attempt to point out the logical way in which that "more", once its existence is accepted, may operate within our brain. One can only believe it, or not.

# 2. Structural and Functional Properties

## 2.1 The Brain Mass

My purpose in this chapter is to describe the relevant properties by which the brain can be treated in the same way as every other physical object or system. In humans, the brain is a compact mass weighing around 1500 g, a mass which has progressively increased in the course of phylogenetic development. However, there is a wide, unfilled gap between the brain mass values of humans and those of the higher present-day primates (the brain of the largest ape, the male gorilla, is about one half the size of the human one) hinting that the mass problem is not irrelevant, since the human brain should, thus, contain at least twice as many nerve cells, or "neurons".

Only a little more than one century ago, the fine structure of the central nervous system (CNS) began to be routinely subjected to investigation, after its basic constituent, the neuron, was recognized as an entity. Actually, the mass of the brain is made up not only of cell bodies, but also of a highly developed felt of cell branchings, for each neuron one neurite (or nerve fiber) and a fine plexus of numerous branches, tree-like and spread out, the dendrites. The neurite is specialized for conducting nerve signals away from the cells, while the dendrites and the cell body itself act as the receivers of signals brought to them through the neurites of other neurons, which cover with their specialized endings, the synapses, the membrane of the dendrites and of the cell body. The number of neurons ranges in the ten thousand millions, although a near exact count seems to be extremely difficult, due to their uneven distribution throughout the CNS and their different sizes in different places. The neurons are grouped in structures called nuclei in the inner mass of the brain, and distributed in regular layers in the superficial part, the cortex, of the cerebrum, of the cerebellum and of the hippocampus.

Surrounding each neuron and occupying much of the extracellular space (there is some belief that it is the whole extracellular space) are other cells, the glial cells, also numbering several thousand millions. Their function is still somewhat uncertain, though we can see that there are several kinds of them. These cells, unlike neurons, do not posses an apparatus for transmitting signals (at least not the same kind of signals as the nerve impulses), but are certainly interacting in some way with the neurons, particularly if we accept the view that they represent much of the extraneural space, probably by interfering with the ion distribution on both sides of the neural membrane (which is the basis for excitability and thus of the generation of the nerve signals). The uncertainties about the functional role of the glial cells are blurring many pages of the story of the brain we are just beginning to read together, but much of the matter could, perhaps, find its logical place after the recent discoveries about the chemical organization of neural activities (see below).

The number of neurons does not increase after birth, so the increase in the brain mass is a matter of increase of the neuronal branches (which at birth are just beginning to sprout in most cells) and of the glia. This fact is relevant, as we shall see, to the processes of maturation and learning (see Chapt. 3). Early in the course of life, neurons are actually beginning to decline in number (a loss that cannot be attributed to the effects of senescence), being replaced by the glia. Simultaneously, changes in brain performance begin to occur, detectable at first only through suitable tests, later on becoming evident to everyone. Much later, as a consequence of ageing processes, greater neuronal losses occur, but none of the fundamental brain functions is altogether abolished, probably because the damage (the neuronal loss), although widespread, is nowhere concentrated in either space or time; whereas, a deficit in the functions becomes generally apparent whenever damage is concentrated, even in a small area and in a short time span. This would point to the conclusion that each function has at its own disposal a redundant amount of neurons and that, if time is allowed, many of the deficits of function can be overcome. This property may be included in the rather vague "plasticity", a term used by the neurophysiologists, and indicating here that the remaining neurons may take on somewhat different functions, an event observed more clearly earlier than later in life. We could also speculate, and not without basis, that the precocious decline of the neurons in our brain is the ultimate manifestation of a process of ontological evolution, by which newly-organized functional patterns progressively replace more primitive and earlier-developed ones, a process, as we shall see, very akin to plasticity and to learning (which we shall treat later on). Ageing, in contrast, intervenes much later in life and starts setting unforeseeble limits to function, learning, and plasticity.

## 2.2 Structures and Functions

It is implicit in what was said in the preceding section, that the neurons are the operators of the functions of the brain, and that, in order to carry this out, different parts of the brain are involved in specialized functions. There is a classical anatomical and physiological scheme of specialization of the brain parts in "sensory" and in "motor" sections, their functions being integrated through a host of "associative" structures. This division of anatomically defined structures, to which corresponds a functional counterpart, is present in all brains, even those of non-mammals, such as birds, reptiles, fish, and amphibians, thus revealing what would appear to be a universal, primordial principle of organization, that of functional localization. In this way, all the beings, though enormously different as to anatomical construction, are provided with basic and similar brain structures with their intrinsic functions, so that also in smaller and lower animals we may find, for instance, exact and fine movements as in humans. The swimming or the jumping of a frog or its ability to catch flies with its tongue are movements as precise as those of a monkey gently picking berries from shrubs with its fingertips, or those of Rubinstein playing a sonata. The difference is that jumping and fly-catching practically constitute the whole movement capacity of a frog, whereas a pianist is capable of more diversified and sophisticated movements than a monkey. (A pianist may, eventually, also acquire fame as fly-catcher, but surely no frog or monkey will ever be able to play Beethoven.) The superior motor performances of the human being, as compared to animals, have become possible for the much larger number of neurons involved in motor control in all its "motor centers", but, as we shall see later on, it is not simply a matter of numerical differences.

An analogous situation has developed as to the sensory functions of the brain. Even insects can perceive different hues in the flower colours, perhaps also better than humans, and dogs can follow a scent in the ground, certainly better than any human can do; eagles have "eagle" eyes, a nocturnal animal has no difficulty prowling in darkness and catch a prey, bats sense also the ultrasounds, to which man is deaf, and there are many more animal performances that man is unable to produce at the same standard as animals do. Nonetheless, man has enormously more "complete" contacts with the outside world than any other living being. For instance, particularly developed in comparison with others is the fineness of the nerve endings on his body surface. The senses of the soft skin of his hands and face so clearly distinguish man from other animals that St. Thomas took that feature as being distinctive of a superior intellectual power (as he says recalling the words of Aristotle: "molles carne bene aptos mente videmus", i. e., those who have soft bodies exhibit better mental aptitudes). What it was not possible to know in Antiquity and in the Middle Ages is that in humans the sense organs are backed by an abundance and an intricacy of neural circuits in the brain, unmatched in any other being, and that the widely articulated neuronal complexes on the surface and in the depths of the brain (corresponding to the peripherally highly developed sense organs of the body surface) are strictly interconnected with those subserving all the other senses.

### 2.2.1 Speech Characterizes Humans

The gap I pointed out at the beginning between the volume mass of the brains of humans and of primates is principally due to the development (or should I say the appearance?) of speech, which represents, as we shall see, the paramount, discriminating difference between humans and all the lower-than-human beings. In point of fact, speech does not consist simply of producing definite sounds: the reason why it has brought with it an enormous increase of the brain mass is that it does not involve only the corresponding organization of a pre-existing pool of neurons (which controlled the motor acts) necessary for uttering words. Speech has specifically required new structures, on the cortical surface as well as in the depth of the brain, essential for the function, so that their destruction may involve a severe impairment (as one can see daily in the case of cerebral strokes), and has "added" to the neurons controlling the motility. Also on the sensory and associative side new structures have been developed: whole areas in the cerebral cortex have the function, principally, of making speech understandable. The latter function, though, is not sensitive, because in its essence it implies an act subsequent to hearing, the recognition of the spoken word, as well as of the abstract, symbolized representation of the content of speech, as in written words or drawings. Most of the speech structures have developed out of the brain areas subserving the necessary co-ordination between the sensory and the motor functions.

The quite particular stand of the speech function in the human brain is due to the fact that at its origin, as a first act, speech must be "thought", an act requiring the use of an internal language with symbols (which can be outwardly expressed through the voice, or the drawn and written signs). Thought and speech have grown together in man, and no brute possesses them. A parrot, and other birds too, can be taught to talk, but it is simply learned vocalization. A chimpanzee, with its screeching and grunting and wailing, can express more differentiated feelings than a cat with its mewing, but still these are only inarticulate sounds, exclusively urged by internally or externally aroused sensations, not signs of autonomous thought. The more recent developments in the behavioral sciences could only show extremely scarce abstracting power in non-human brains, and failed to produce any convincing evidence of autonomous "thinking" processes, even in the primates. What in the latter has been called thinking should be interpreted more correctly as an association of memories evoked by the pressure of immediate urges, that is, a phenomenon thoroughly dependent upon the contingent situations in the external world or in the internal organs, not appearing as an autonomous act. In the everyday life of man, instead, this autonomous act is facilitated by the very absence of contingencies from within and without the brain. Therefore, the true origin of speech resides in the brain's capacity for thinking, i. e. of developing by itself its own abstract symbols and of making logical constructs out of them, also independently from what is going on in the outside world, of producing concepts and ideas to which the brain may give a physical expression (but not necessarily) through motor acts, speech or drawing. Physical speech, thus, is not a self-supporting or self-sufficient functional property, it is only the outwardly directed sign of an internal process by which the brain can talk with its "conscious self". Although these concepts may raise opposition on several grounds, I still hold firm to them, as the ones stating the true, clear-cut distinction between humans and non-human beings (see also the Appendix).

## 2.3 Connectivity and Integration

The more we consider the structure of the higher brains, like the human one, the more we become aware that the classical functional distinction between sensory, motor and associative parts cannot be rigidly maintained. Speech and thought demonstrate that the distinction has, perhaps, no reality beyond a semantic usefulness (whenever one has to deal practically with this or that particular brain structure).

All brain structures are interconnected by an enormous abundance of anatomical links, on which they base their potential for extended, reciprocal interactions. Here, interaction simply means that the activity of one structure may modify that of a large number of others (and of course be modified by many), the basis of that process the neurophysiologists call "integration". Let's look more closely into this reciprocal interaction. As I have stated in Chap. 1, the properties of the brain are founded on the detailed anatomical organization of the structure, but the latter provides only the physical, or material, substrate through which the functional characteristics have developed, whereas our attention is oriented towards the functional properties of the structure, leaving the anatomical details in the shadows.

The ten thousand million neurons of the human brain keep in touch with each other through the numerous branchings of their neurites, of which there are sometimes thousands for each neuron. If one assumes a conservative figure of a few nerve signals per second for each neuron (all being actually active), we must account for billions of nerve signals each second going to and fro in the maze of neuronal circuits throughout the brain. This makes our brain a giant system of communication, miniaturized in the small volume presently occupied, the most complex living system known so far. It is obvious that the problem of obtaining a "blue print" of the circuitry in the brain should be of paramount importance for everyone concerned with its functions. We are now in possession of a map (though quite incomplete, we must say) of at least the principal connections pertaining to large functional classes, such as sensations through the special senses, motility, sleep-arousal, regulation of autonomic functions, emotions, etc. However, the more one proceeds, the more complex is the resulting picture, since within any of the anatomically defined groups of neurons the individual cells do not seem to be endowed with exactly the same properties. In other words, they do not "do" the same things, and their activity contributes to more than one function. Thus, one can never be sure whether this happens because of a real difference in the intrinsic properties of the neuron, or because of a difference in the individual connections which results, finally, in a difference of the signals being received and processed. Although morphological studies show us differently shaped cells (as, for instance, in the different layers of the cerebral cortex), there is little evidence that different neurons have essentially different intrinsic capacities. The differences in the relative development of the cell body volume and of its branchings, neurite and dendrites, do not correspond to differences in the basic mechanisms by which a neuron receives, processes, and sends signals. Rather, they seem to indicate the different importance of groups of signals arriving at the different regions of the neuron's surface, and the different type of participation, or role, of that neuron in the complex of the structure it belongs to. The conclusion is that it is not only the blue print of the circuitry which eludes us, but also the characteristics of the signals in relation to the functional role of each neuron.

Even assuming some degree of functional homogeneity of the intrinsic neuronal properties within the same anatomical group (we shall discuss at length the problem of homogeneity in Chap. 5), knowledge of the exact characteristics of the signals received appears as insoluble a problem as that of the complete blue print. The reason resides in the large number of nerve endings on every neuron, which sometimes literally cover the whole surface of the dendrites and of the cell body, and steadily transmit signals from neurons located in the most different parts of the nervous system. Only in particular situations, where it is possible to control the generation of the signals, is it also possible to follow the latter up to the end of a pathway in a brain center, as in the case of sense organ stimulation. The latter is a branch of neurophysiology which has been particularly investigated, because together with the immediate interest in knowledge of the function of a sensory organ, it is the most suitable approach for studying the general relationships between the membrane potential changes of nervous cells and the ensuing nerve signal discharge. Locating the neurons to which these nervous signals are directed requires some additional ingenuity, but once the technical difficulties have been overcome, one has all the means for studying the effects of known signals on single neurons. In the vast majority of cases, however, complete knowledge of the origin of the arriving signals and their pattern of distribution in space (i. e. on the neuron's membrane) and in time cannot be obtained to any extent, thus barring a particularly useful line of investigation. This impossibility has constrained the neurophysiologists to look for other methods: in general, they have resorted to treating the activity of a neuron (which is at every moment the necessary result of the immediately preceding, though unknown, pattern of signals received) rather as the outcome of a statistical process. Out of this, which can at most be considered a second best choice, many neurophysiologists have created a tenet, i. e. the principle that neural activity actually "is" the result of a stochastic process which can be studied only statistically, thus practically assuming that knowledge of the exact structure of the signals received is not relevant (which has not at all been demonstrated).

At any rate, it is true that the investigator in most cases has to content himself with knowing what is connected with what within a restricted subsystem, the simplest possible of the wiring plans: he eventually classifies the neurons as belonging to this or that "center", and, within a center, classifies them on the basis of their responses to known stimuli. But all the painstaking study of the connectivity within the nervous system has finally revealed that the grouping of neurons in centers or subsystems, and the grouping of these in systems, is not the ultimate answer to our needs, for the very simple reason that if, besides the direct anatomical links between the neurons of any two different structures, we consider also the indirect ones, i. e. those through chains of intermediate neurons, then the possibility for each neuron to communicate with any other through whatever complex circuitry is extremely high. Without resorting to the famous, destructive sentence that "everything is connected with everything else", nonetheless it is true that the activity of any center whatever may influence in the long run, to a larger or smaller degree, widespread areas of the brain, if not the whole of it.

If we accept all the above, we must also conclude that one single structure alone cannot be responsible for a function, because the widespread connections with many others make each structure a participant, to some extent, in every neural activity. Therefore, we must further discuss the significance we attach to the term "function". In view of the widespread connectivity between different subsystems, it is hard to accept the nervous functions as pure "categories", for instance, motility, sensation, etc. At most, for instance in the case of sensation, one could restrict to this pure category those neural events evoked in a subsystem by the signals originating from a sense organ, and reaching it through the shortest nerve path, that is, through the straightest possible connections. This would arbitrarily restrict a "pure" sensory function to the neural events temporally closest to the one which originated them, but would also divest those events of the significance we generally give to the term function. The resulting concept would now seem to be a mutilated abstraction, represented only by the initial stage of a process which, instead, normally proceeds with the arrival of signals from other subsystems, and completes an operation only initiated by those which arrived earlier through the shortest path. Without the attendance of all the different sequences of impulses, it would become altogether impossible to define a function whatever, since no function is contained in that tiny interval of time (of the order of a few milliseconds) between the arrival of the first signal through the shortest path and those through the next shortest. The necessity for the earlier and later arriving signals when defining a function is, again, best exemplified in the case of sensation. The object of sensation has many properties, e.g. form, volume, roughness, temperature, and also colour, luminosity, or maybe frequency, pitch, intensity, duration, and so on. The signalling of many of the properties of the object of sensation occurs, in general, through different nerve routes, along which the nerve signals are travelling at different speeds. Therefore, even starting at the same time in the same place, the signals always arrive at the next place scattered at different intervals. Furthermore, the simple arrival of the signals at the first nervous station beyond the sense organs does not by itself constitute a sensation. No matter how short-lived the stimulus it has aroused, the nature of an object must be recognized, a brain process requiring the analysis of its properties from the signals received, and their synthesis in a judgement, with memory comparison etc. Moreover, all this is also accompanied, not followed, by motor acts (maybe only a twitch or an increase in the tension of some isolated muscle), emotional and autonomic reactions. The accompanying nervous events are produced by the same sensory signals which course in collateral branches to other nervous structures, spreading the sensory information over wide brain areas and setting the basis for generalized reactions. It is true that lesions placed in some key point, or destructions of large portions of nervous tissue can somehow isolate the "sensing" from its accompanying phenomena, but is also true that what remains is no longer a "normal" sensation. The more the sensory event is deprived of its attributes, the more the conscious brain becomes unable to judge correctly the nature of the sensation, in the extreme, to even attribute the sensing to the correct modality.

A closely related question concerns the general way of processing the information, i.e. either through serial or parallel circuitry, in other words, whether the processing is performed in subsequent, well-defined steps or simultaneously. In the field of sensation, the first modality requires that sensing, analysis, synthesis, memory comparison, recognition, emotional and autonomic reactions are all taking place in successive steps (as well as in different structures) following some logical sequence which might not be the one which I have, exempli gratia, jotted down. The second modality assumes that all the steps, which may be tentatively called "partial functions", are taking place more or less simultaneously, also being required in this case to occur in different structures. In the first case, there would be an ideal chain of events, which has not been demonstrated to exist in such a way, there being no evidence in favour of an exact temporally defined sequence of sensory events, while electrophysiological as well as psychological data have revealed that some processes which have been assumed to occur at later stages may actually intervene earlier, shaping nervous events in the first nervous stations. However, both modalities, serial and parallel processing, seem to suggest the same conclusion, i.e. that nervous functions do not exist in nature as pure, isolated categories of events in circumscribed areas, but must necessarily result from the interplay of a wide variety of structures at the same time, and for some definite interval of time (this is a concept I shall further expound in later chapters).

Obviously, a distinction between functional categories must be maintained for practical purposes, in investigation and teaching (as semantics needs to fractionate the discourse in a sentence), but one never sufficiently stresses the fact that nerve function has two aspects: spatial, as it involves large and widely spaced structures, and temporal, as it develops in a specific interval of time and loses significance if considered over a time interval which is not "natural" for it. In other words, the essence of any function is "integration". In neurophysiology this word does not have, in general, the exact significance it has in mathematics, although conveying the same philosophical meaning, that of unifying at some higher level, of transforming local events into large-scale phenomena, in our terminology, making the "partial" functions, such as sensing, analyzing, memorizing, recalling, abstracting, symbolizing, moving, secreting, etc., all become acts in the function of something, of some unifying function, call it perception, awareness, logical thinking etc. I am fully aware that I am changing the common intuitive conception of function, but I am thoroughly convinced that the brain's operations are to be taken as a whole, that there is no brain operation due to only one of its parts, that the brain does not operate by parts, but as a whole (see also Chap. 3).

### 2.3.1 Feedback

Among the innumerable patterns of interconnections between neurons, one has particular importance because of its generalized role in operations, "feedback". In terms of the brain's structure, feedback indicates an anatomical path which returns to its point of origin in two possible ways: either through directly "recurrent" collateral fibers, branching from a neurite and coming back to the same neuron from which the neurite originated (a situation rather specific and peculiar to some structures); or through longer circuits, in which many cells in a chain (generally belonging to different structures) are interposed. The latter seems to represent the by far more frequent occurrence. Leaving aside the details of the anatomical substrate, feedback is a principle of operation, representing the means by which groups of neurons or subsystems tend to work as a unit, since the neural origin of an event becomes the receiver of the effects originated by that same event, and gives rise in turn to other neural events. We must be aware that this closed loop of nerve messages is obliterating the very concept of "origin", since we can soon talk only of events "circulating" in the feedback loop. This is generally ignored, but needs to be stressed: each neuron of the feedback circuit is simultaneously the source and receiver of information in a closed circle, and as we shall see next, the information may be different at every link of the chain.

As a principle of operation, feedback reminds us of the man-made mechanisms for automatic control of operation in physical systems. It represents the basic principle in cybernetics, where it conveys the significance of information fed back to the controller, to keep it updated about the state of the controlled system. The same concept could be accepted, probably without any change of significance, for the nervous system, although we have to clarify some of the peculiar aspects of neural feedback. Cybernetic problems also exist in the nervous system, e.g. the control of muscle tonus in posture, the maintenance of body balance in space, movements to reach a target, all automatic movements, the maintenance of a steady level of blood pressure and of respiratory rate, etc. However, the feedback loop arrangement of nervous connections is ubiquitous in the brain, and thus seems everywhere to involve the same principle of operation, well beyond the trivial and limited cases in which some physical action must be controlled. I can see the meaning of this ubiquitous presence in the necessity of creating a unity (as in cybernetics also) where the interruption of the flux of information brings about the disappearance of its self-sufficiency or, in other words, of its "functional" unity. If, in some cybernetical subsystems of the brain, feedback assures unity with regard to a physical operation performed in, or by the body, in the great majority of subsystems it must assure unity at some "higher" level of operation.

The fact that in the feedback loop there may be inhibitory neurons (neurons which can decrease or block the activity of others) gives to many of these loops a particular function. It is, in general, a property of the shortest of them, with only one neuron in the feedback path, the inhibitory one closing the chain. Their general function is believed (with some certainty, since there is enough experimental evidence to support it) to be the maintenance of a fairly stable level of activity within the chain, perhaps also to act as a way of preventing the neurons from becoming exhausted by generating signals in high numbers and at a high rate (as may occur, for instance, in response to strong sensory stimuli). In some circuits, temporally programmed inhibition through short paths may act as an "on-off" switch of the neuron's activity. There also seem to be excitatory feedback loops of this short type, but only scarce evidence of their existence is available, although it is clear that they could maintain their general stabilizing character in circuits where strong inhibition is otherwise produced.

The overall significance of the short feedback loops seems to be quantitative. The shortest of them carries exactly the same type of information issued from the cell body of the neuron, thus, as a consequence of the closed loop, there cannot be any immediately subsequent change in the quality of the information sent out by the neuron. In other words, it would always be the same information recirculating, scaled up or down. Due to the inhibitory link, an "all-or-none" effect (the switch effect) may also result. Things would seem to be different for longer, multineuronal feedback circuits. The latter involve neurons belonging to different structures, generally also being part of other feedback loops, in which every neural message arriving is presumably processed with some changes in the quality of the information contained, since upon every structure there is convergence of signals from different sources. Thus, the action of the long feedback loops would be more dependent upon the physiological significance of the signals at every step within the chain. More generally, these feedbacks seem to maintain a logical, sequential order of the neural operations performed by each structure of the loop, in which the informational content is progressively defined and refined up to a point at which the processing may be considered terminated, or has reached a steady state.

Although we can transfer to the nervous system the concept of feedback as the basis of control systems, it is also true that in the brain this type of control, which implies automatic operation, is limited to only a few partial functions, mostly involved in the control of the automatic aspects of motility and in the vegetative sphere. In other cases, the concept of control is blurred to a vanishing point, and the unity in these circuits is related to such functions as adaptation, decision, choice, comparison etc., in which feedback is no longer the unique, basic, functional factor, but is only one of the many elements involved in a complex function.

## 2.4 The Neurochemical Aspects of Brain Organization

The picture we have obtained so far of our brain is that of a system in which the fine and detailed anatomical organization of connections is certainly the basis for the wide gamut of its functions, and at the same time a system in which widespread cross-connections and feedback loops make possible largely integrated operations. As previously stated, this has generated the conviction, based after all on everyday experience, that in every occasion the brain participates as a whole, with all its subsystems, in every function, obviously in a very differentiated way as far as the participation of individual subsystems is concerned.

To strengthen this belief in largely unitary processes, neurochemistry has recently contributed evidence of the secretory capacities of nervous tissue, which constitutes the most exciting discovery of the last decade. It has been known for a long time that a part of the brain, the hypothalamus, secretes peptide hormones, which course along nerve fibers and are subsequently released into the blood stream by the posterior lobe of the pituitary gland, having diffuse bodily effects. Since the hypothalamus also controls important body functions (such as water balance and food intake, metabolism, body temperature, etc.) it was more than natural to look for other secretions controlling these functions. This led to the discovery that hypothalamic nuclei produce other peptidergic molecules, the so-called releasing factors, which induce the release of hormones in the anterior lobe of the pituitary. Sensing a promising harvest, the neurochemists have proceeded to isolate from nervous tissue more molecules with effects on peripheral organs (such as the digestive tract), but also on the brain itself. Moreover, molecules produced in the periphery seem to have the same structure as those produced in the brain. Thus, the brain seems to possess two means for communicating with peripheral organs, a) nerve fibers belonging to the autonomic nervous system, largely derived from the hypothalamus itself, and b) chemical messengers. Both are two-way communicating systems, afferent and efferent nerve fibers, to and fro molecules.

The discoveries are far from being complete, so that no one can yet foresee their future developments, or their exact bearing on the knowledge already available about the brain's operations. But, at least this much can be said: if the brain produces molecules influencing its own activity, even if not of all its parts, then the brain has the means of controlling its own operations. As I was saying in the previous section, the brain is operating as an autonomous unity, autonomous in that its control would necessarily depend upon what is going on in its own neuronal circuits.

Some of the peptides secreted by the brain exert a very restricted influence, or at least are more easily detected in restricted areas, such as the so-called "endorphins" (molecules with a morphine-like action), the most striking and popular of the more complex peptides of the brain, which probably intervene in the still-obscure process of pain and pleasure sensation. The action of the endorphins can be explained by their chemical affinity for "receptors" on the neuron membrane with which the drug morphine can also bind. This line of research is most promising as it provides detailed knowledge of the operations of particular groups of neurons, and suggests the physiological basis for explaining many a pathological phenomenon of the nervous system.

There is also strongly renewed attention of physiologists and chemists to the "mediators" of nerve signals, molecules which normally provide for the transmission of nerve signals from the terminal endings of a nerve fiber to the membrane of another neuron within the specialized points of contact called "synapses" (see also the next section on nerve impulses). The definition of the synaptic mechanisms for signal transmission is one of the cornerstones of neurophysiology. With reference to it, the newly discovered molecules seem to be of two categories, those having well-identified receptors on the postsynaptic membrane, as the mediators do, thus appearing to be very specific to some classes of neurons, and those which are apparently not specific, since receptors for them have been discovered outside the synaptic contacts. In the first instance, peptides characterize subsystems of neurons chemically defined by their specific synaptic receptors. In the second instance, the peptide action is much more generalized, although interfering in an unknown way with the activity of the neurons, most probably acting on not only the membranes, but on the metabolism of the cells as well. As for the more specific mechanism, there is also convincing evidence of molecular fluxes along the neurites from the cell body to the synaptic endings, and of the existence of a larger number of mediators than has been assumed so far. In other words, there exist different "chemical subsystems" of neurons of which we are now obtaining the first "maps".

The main characteristic of the chemical maps is that they do not necessarily or exactly coincide with the anatomically defined systems of neurons, hitherto painstakingly outlined with refined tracing methods. It is still too early to make predictions on the true extent of an anatomical-chemical dualism and on the full impact of these new data upon the problems touched upon in this book. Among the data gathered so far, only a few lend themselves to general comments. Within an anatomically defined system of neurons there are subsystems endowed with differentiated chemical properties, as defined not only by their synaptic mediators but also by the membrane receptors. The subsystems all throughout the brain, composed of cells with similar chemical, synaptical characteristics, can then be reached by one chemical controller, common to all of them, an operation which seems to be carried out regularly within the nervous system, providing the means for influencing widespread, though selected, areas of the brain. The other, more diffuse chemical control is not a typical synaptic mechanism, rather a kind of "modulator" of synaptic operations, by which the nerve impulse output of neurons may be either increased or decreased. Certainly, the latter is not the right way to describe the action of the modulators, because as mechanisms controlling clusters of subsystems, their operation should be described in terms of the functions subserved by these structures.

It is easy to see how neurochemists and pharmacologists are jointly involved in this line of research: in these last few years systematical physiologically-oriented therapy of the nervous system has become possible, the beginning of a fruitful collaboration for all concerned.

Neurochemistry is lending further support to our initial assumption. The principles of our brain's operations, albeit based on its structure, cannot be exclusively explained by the structure: the latter is used in the brain as bricks and beams are used to build a house, but the very existence of the house and its many functions are made possible only by the laws of engineering, of architecture, and of assembly, through which the works are organized. From now on, then, we shall confine ourselves to this latter aspect.

## 2.5 Nerve Impulses and Signals

We have been talking about nerve signals without explaining the significance of the term, assuming only that within the nervous system communication between neurons is maintained by nerve events – signals, or messages, coursing along the nerve fibers. It befits this chapter to be concerned with these means of communication. What we shall henceforth call "nerve impulses" are the elements building the signals, or messages, sent from a given neuron to many others, through the manifold branchings of its neurite. Their presence in the complex circuitry of the central nervous system is, thus, as fundamental as the specific structure of the neurons and of the anatomical organization of the circuits, since the very existence of a network of neurons is justified by their being senders and receivers of nerve impulses. Even if the latter are by no means the sole representatives of neuron activity, nonetheless they constitute the most relevant sign of it so far. So, to build a theory of brain function, we cannot omit a short description of the characteristics of the signals.

We shall not bother much about the physico-chemical nature of the nerve impulse, although its biophysical aspects have long been, and still are, objects of investigation. The nerve impulse is said to be conducted along the membrane of the nerve fiber as a quickly reversible disturbance of the asymmetrical ion distribution between the internal and external media of the fiber (maintained in equilibrium by the expense of metabolic energy). The impulse is not a process actually "travelling" along the nerve fiber, rather, it is generated at each successive point by the presence of that disturbance of ionic equilibrium at an immediately preceding point. Considering its main electrical attributes, the potential changes across the membrane, the nerve impulse maintains all along, unchanged in an all-or-none way, its physical properties, i. e. amplitude and duration, up to the ending in the synapses, where it induces the liberation of the mediator molecules in quantal amounts. The latter, in turn, set up in the postsynaptic membrane another ionic disturbance, eventually propagating to other parts of the neuron, and finally giving rise to another nerve impulse in the corresponding neurite. Generally, many impulses must impinge on the postsynaptic membrane in order that a new impulse might be generated in the postsynaptic neuron. Suitable summation of the successively incoming impulses is necessary, both in space (over the surface of the membrane) and in time. (This is merely one of the integrating operations of the neuronal membrane).

The significance of all this is that a single impulse cannot constitute "the" signal received by a neuron, the actual signal being made up of a variable number of impulses, and it is the general belief (although verified only in the spinal motor neurons and in peripheral sensory structures) that a signal is made up of a particular sequence, or time pattern, of impulses. Only the patterning of the impulse sequence can define the signal "structure" in a nerve fiber, since the amplitude and duration of the individual impulses are constant for each nerve fiber. Things are somewhat different at the neuronal membrane, where the signal acquires a spatial dimension through the confluence of several different nerve fibers with their endings, each with its particular time pattern of impulses. However, the impossibility actually knowing exactly the spatial and temporal distribution of the impulses at each of the numerous synaptic endings prevents the definition of the exact structure of the signals received by each neuron.

What I have called integration by the neuronal membrane has not simply a quantitative aspect, because the different origins of the impulses converging on it confer to the operation a qualitative significance, which could be exactly known only if the significance of the signals conducted by each individual nerve fiber could be known as well. To complicate the matter, there are synaptic mediators which inhibit the ionic changes set up in other synapses of the same cell, i. e. mediators able to reduce or altogether abolish the excitation induced through excitatory synapses. This means that in the structure of a signal generated by a neuron there are pauses, or other changes of the time patterning of the impulses, linked to inhibitory effects. Again, this introduces into the operation performed by the membrane another qualitative change, not simply a quantitative one, as it might appear at first sight, consequent to what was said before about spatial integration, since the inhibitory synaptic endings belong, as a rule, to different types of afferent fibers. Recent evidence, however, weakens the assumption of a strict separation between excitatory and inhibitory nerve endings as belonging to different types of neurons, since it has been demonstrated that it may be the receptors on the pre- and postsynaptic membrane which define excitation or inhibition for one and the same synaptic mediator.

Thus, together with the anatomical connections and with the chemical characteristics of the neural net, the signal structure determines the characteristics of the system of communication between the neurons. So far, however, the way in which the structure of the signals is determined in the nervous system is unknown, all evidence about signal structure being still from the sense organs. In the latter, an applied stimulus can be exactly calibrated as to, for instance, rise and decay times, peak intensity, steady state intensity (if there is any), overall duration, etc. All these parameters are translated via trains of impulses (the sensory signal to the centers), which have characteristics varying with the parameters of the stimulus. In this way, the essentials of the signal structure have been found to be the instantaneous frequency (the reciprocal of the time interval between two impulses), the average frequency at steady state, the overall number of impulses, the duration of the impulse train, etc. In the central fibers, though, the signals never maintain the same characteristics throughout, because at each synaptic station (i. e. at each neuron) the structure of the outgoing signal is shaped according to the coincidental arrival of other signals from different origins.

The problem, in its essence, is that of the "coding" of the signals in the nervous system. From the sense organs, which transduce the quantities referrable to the parameters of the stimulus into sequences of nerve impulses, the signal is coded in such a way as to maintain the quantitative information. Analogous quantitative information must be contained in the impulse sequences building the signals sent by the motor neurons to the muscles, since they must be coded in such a way that the contraction produced is quantitatively defined as to strength, velocity, duration, etc. However, outside the phenomena occurring at the very input of the information from, and the the output to, the physical world, all the other nerve events in the central structures have not such an exclusive quantitative significance. Certainly, in the large majority of cases, the informational content won't have a quantitative character at all, and the code, accordingly, ought to be different. These problems will be approached again from other points of view in other chapters.

By "qualitative" we are actually indicating the time patterning and all those characteristics of an impulse discharge which make from it a signal with a significance (and an "effect") different from others. "Quantitative", on the other hand, can only refer to the numbers represented in the impulse sequences through which physical quantities are coded (obviously, a scaling up or down of all the numbers would retain the signal's precise quantitative significance). If, simultaneously, there also occurs a modification of the pattern's characteristics, the change becomes a qualitative one: this concept needs to be clear in relation to the induced changes in the overall level of the activity of the neurons, as we shall see in the next chapters. A change of level can be quantitative only if the characteristics of the discharge are not modified; the latter, however, is a very likely event because the time intervals between the impulses and the duration of the trains are, in general, not simply scaled up or down. For a change in the activity level, then, a purely quantitative significance can be assumed only if it can be demonstrated to occur simultaneously through a re-scaling, or better, a transformation of the reference coordinates of the whole system. This shall be discussed at length in the last four chapters.

## 2.6 What Shall We Substitute for Structure?

It has repeatedly been made clear in this chapter that I am not basing a general theory of the brain function on the specifications of its anatomical structure. It is practically impossible to obtain a complete blue-print of the circuitry, but it is also now apparent that we cannot obtain detailed knowledge of the signal's structure either. What will represent the bricks and the beams of the edifice of our theory must be independent of anatomy and of the signal's structures, and must be a substitute for them. Let me just hint, closing this chapter, what the substitutes will be. A long road is still ahead of us because several possible theoretical approaches will have to be examined and partially discarded, but I must give the reader some idea of the destination we are heading for.

I am introducing here, for the first time, the problem of the "forces" within the central nervous system, a term I am not using as an analogy, but in its full physical significance (the problem, however, is to be repeatedly discussed in Chaps. 6, 8 and 10). As we have seen, our brain is a system made of innumerable parts, largely interconnected through both nerve fibers and chemicals. Now, what are defined as connections are only the means by which an "influence" of some sort is transferred from one place to another. Of course, the signals are the carriers of this influence. However, in them we can see two aspects, i. e. the informational content, surely, but also something we can consider as a "force", which constrains the receiving neuron into a different "type" or a different "level" of activity, or both. The set of anatomical connections between the neurons might be conceived of as building up systems in which there are short-range forces, to use the terminology of physics, in as much as they are exerted at the contact points between the neurons, the synapses, binding parts which are, thus, constrained to a kind of common life (like the subatomic particles of an atom). But in the nervous system there are also what we might call long-range forces, those which, seemingly, do not need direct material contacts between two bodies, exerting however an influence profoundly modifying their state. These are not transmitted through empty spaces, rather through the blood stream and the interstice, and would be represented by the "modulators" (perhaps, also by field effects). We must say, however, that the two-forces concept was already common in neurophysiology long before the discovery of the chemical modulators: actually, in every nerve signal there is this two-forces aspect. The information contained, as translated in the patterning of the impulses, can be considered as short-range force, since it forces the receiving neuron to act according to the information of its input, so that in the output there will be always a trace of that information. (In other words, the output will always exhibit some specific correlation to the input.) On the other hand, the mere impulse content of the signal, the number of impulses of which it is made, independently of the information carried, is more akin to a long-range force, since it tends to produce a nonspecific change of activity (for instance, of the average level) as the modulators are actually supposed to do. Thus, the first type of force has some specific meaning for the neuron on which it is exerted, whereas the second type does not, determing only the conditions under which the neuron performs its functions. The latter type seems more characteristic of particular subsystems of the brain, and might be viewed, as we shall see, as the method of rescaling or of coordinate transformation (a problem debated at length in Chaps. 9 and 10).

Once we have recognized the presence of forces, we must also accept that we shall have to consider "fields" of force. The more obvious of these are established through the electrical changes (the membrane charge unbalancing) of the thousand millions of neurons and of the thousand more of nerve endings with their continuous activity. Of course, locally, these fields are quickly varying in strength and orientation, thus mutually compensating in normal conditions, but they are known to assume definite orientations and considerable intensities in particular states of the brain (epilepsy, repetitive sensory stimulation, drug actions, sleep) thereby influencing the neuronal activity, although direct, formal proof of this field influence is still scant. It is obvious that we are not concerned with this type of forces and of fields; what we are interested in, is not related to the electrical aspects of the nerve impulses per se, rather to what they represent in terms of function, thus, our forces and fields are not to be measured in electromagnetic units. The useful concept of fields of force will be resumed and extended later on, when we shall consider the brain as a "continuum". The reader may now ask what do we gain by substituting forces and fields of force for anatomical details and signals' structure knowledge: well, the forces can be measured and physically treated, and their fields build "functional structures" which may form the common basis for every consideration in the domain of neuroscience, that's all.

# 3. The Key Property: Organization

## 3.1 Brain Functions Are Unitary Processes

This concept has already been outlined in the previous chapter. Here, I shall try to expound it better and to discuss it to some extent. It has been made clear that the main reason why the human brain outclasses those of other animals is the development of the structures by which language, internal and external, is made possible. With the increase of complexity, the properties of the anatomical structure are increasingly exploited to advantage by means of "organization". Organization is actually a principle through which the different anatomical structures cooperate ("operate together to a common aim"), so that the properties of each are enhanced by mutually collaborating as an ordered unity.

I shall first try to make clear what the term organization specifically implies for the brain. Even if the organization is founded on the maze of connections between the thousand millions of neurons, constituting the anatomical substrate of organization, it is not simply (if we can use this word when dealing with such a complicated wiring plan) a matter of connections. The superior functional performance of the whole enables man to rise far above and to overshadow all other beings. The human brain is certainly more complex than any other: however, this is not basically the result of the larger number of neurons contained in its volume, rather, of the fact that the addition of more structures and their related connections has rendered possible an increasingly improved "use" of the whole complex. We could picture an imaginary predecessor of ours, slowly ascending the stair of evolution, while phylogenetic development brings to his brain the addition of new structures at each step. That there has been phylogenetic development in the brain has long been recognized and, thus, we have prefixed to some of its parts terms like archi-, paleo- and neo- in order to indicate the outcome of a temporally differentiated brain growth. Coming to the point, the brain of our friend is not simply an ensemble of structures layered on top of each other: every new structure appearing, or developing from those preceding, is not simply added to older ones, rather, it is fused with them. There is a growing together of old and new. Old and new neurons belong to the same anatomical division, old and new fibers connect both old and new structures. The consequence is that we cannot "peel off" the newer structures in the human brain, expecting to progressively reduce it to lower and lower stages of development: by this we shall simply obtain a damaged human brain. Of course, this is also true for every brain of subhuman beings. If we try to better analyze the situation, we realize that we are confronted with a basic property of the brain, which prevents any of its parts from entering into an operation independently of the others. Thus, organization implies that there are parts constituting a whole. However, in its essence, organization is the property making a functional whole out of parts. The conclusion might invite further considerations about the standing of man (as represented by his brain) among the living beings. Man does not differ essentially from animals for the excellence of his digestion or of any of his vegetative functions: it is the excellence of his brain's functions that has made man what he is. The brain's operations are so tightly bound to bodily functions (see Sect. 2.4 of the preceding chapter) that there must also be in the latter a trace of the presence of such an excellent brain. I shall make further comments about this in Sect. 3.4.1, and some more metaphysical considerations in the Appendix.

The assumption that the brain operates as a whole on every occasion might sound a bit too positive to the large majority of neuroscientists, the thought going immediately to those (apparently) very simple nervous events which occur mostly unnoticed, and are almost as unavoidable and automatic as the physical phenomenon from which their name derives: the reflexes. However, even the simplest of our reflexes (which indicate acts, motor or secretory, in response to the stimulation of specific sensory receptors) does not consist only of the contraction of one, or a group of muscles, or of the secretion of a gland; the reflex does not exist, under natural conditions, as a unique, isolated event involving a single effector. The reflex, though comparable to an utterly deterministic phenomenon, is not completely determined by the stimulation of its receptors, rather, it is always part of a functional organization from which it derives its "purposefulness", the very reason for its existence, i. e. its physiological significance, always consistent with the situation, (which may require either potentiation or depression of the reflex, or even its complete suppression if necessary). To achieve all this, on the shortest chain of neurons involved in the reflex stimulation, the "reflex arc", must impinge a variety of signals coming, either directly or indirectly, from different parts of the brain and each contributing its part, to define the correct physiological significance of the reflex arc's activity.

It is true that through the reflexes the brain reaches goals without having to bother "thinking" about them, because for those purposes a simpler nervous organization is sufficient, permitting moreover a fully automatized operation. The reflexes are, so to speak, built-in, or pre-programmed like the hardware of a computer. This does not mean that these reflex arcs operate in isolation from the rest of the brain. No activation of a reflex arc alone would have any meaning, not even those related to the survival of the organism or the maintenance of its integrity, since no successful escape from, or avoidance of danger can be obtained solely from simple reflex arcs. For instance, in a decapitated frog the nearest leg may reach a chemical irritant placed on the skin and, occasionally, may succeed after some attempts in pushing it away. One would think, then, that such a reflex does not need to involve the brain, as if the spinal cord, where the neurons active in the reflex are located, could manage perfectly without the rest of the nervous system; as if they were separated, or simply added, systems, but this is not so. In an intact-brain frog the reflex is not reduced to the elementary acts just outlined, i. e. the latter do not constitute the "real" reflex. In the intact animal the reaction to the same harmful stimulus is consistently more lively, infinitely more precisely goal-directed; it involves all legs, is performed together with other motor acts (partly also of reflex nature, such as those required for the continuous shift of the center of gravity and the consequent counterbalancing of body weight during movement); there are pricking or pain sensations; generally the frog moves away, eventually jumping farther, thus exhibiting whole-body participation. This is the real, purposeful reflex reaction. However, all this is not merely added to the simpler, or skeleton reflex (as produced in the isolated spinal cord) by intervention of the nervous structures above the spinal level. The higher structures by themselves could never produce such an integrated response, since in order to be able to command integrated acts they need continuously correlated sensory information from all the sensory organs. The brain operates only in this way. Assuming otherwise, would be like maintaining that a symphony may consist of playing the theme on a piano with the forefinger, disregarding the full orchestra.

In man, the consequences of a lesion affect normal behavior more profoundly than in animals, and in the case of separation of the spinal cord from the cerebrum, the consequences are catastrophic. After spinal transection, (from tumours or wounds,) if death does not immediately ensue, all spinal cord reflexes disappear. After only a few weeks or months incomplete and ineffectual reflex movements can be produced, but the control of functions normally performed through the spinal cord is lost forever. Other lesions, higher in the brain stem, may bring about a syndrome similar to decerebration (as found in animals) in which the spinal reflexes of posture are exaggerated (the myotatic reflex which provides for contraction counteracting muscular elongation thus counteracting principally the muscle stretch due to gravity). A typical rigidity appears in all four limbs, permanently maintained although there is no stretch to counteract. We are thus in the presence of a reflex without any purpose, an act with no function attached. Actually, we cannot say it is a reflex any more if we still want to retain this term in its true physiological connotation, that of a purposeful act performed automatically by the nervous system.

What happens to reflexes as the consequence of a lesion also happens to the general control of motion as a consequence of anatomical alterations of a different nature (trauma, tumours, degeneration, ageing) in the nuclear complex formerly called "extrapyramidal", in the cerebellum, and in their nervous pathways. Highly distressing, purposeless movements may appear, such as tremors, athetosis, chorea, ballism, as may spasticity, sluggishness, incoordination, ataxia and dysmetria, all strongly interfering with the normal execution of movement. If we maintain that physiological movements as such always have a purpose, a goal to reach, then the consequences of lesions show that in order to preserve a purpose to movement, the integrity of the whole brain must be protected. The basis of the abnormal movements, and of the movement defects is the interruption of the wealth of feedback pathways continuously informing the cortical and the subcortical motor systems of the progression of the movement (through signals from all the sense organs in the muscles, joints, integument, eyes, and vestibulum). All this information is indispensable to adjust the nerve impulse discharges which control the direction, the speed and the force of the moving parts, so that the goal may be reached exactly. This is the best demonstration of what was stated in Sect. 2.3.1, that the feedback principle maintains unity of operation in the brain as a whole.

As we have seen, the rigid separation not only of brain parts, but of functions, such as motion and sensation, does not seem to make much sense. However, at first sight, while to be correctly performed the motor act always needs the participation of sensory functions, the latter would also seem to be conceivable in the absence of motion, as in completely paralysed subjects. Again, this is not true. The acts of sensation affect all brain structures, including the motor ones. As we said previously (Chap. 2), sensation is made up of a multitude of events, which are largely taking place simultaneously along parallel channels, in that sensing an object involves the analysis through the senses of all its qualities (form, solidity, roughness, and temperature, but also colour and hues, eventually tone and pitch, smell and taste, etc.). The analysis is never disjoined from a synthesis, i.e. a simultaneous recognition of the nature and significance of the sensed, from a comparison with previous experiences through the involvement of the memories associated with it, from the emotional counterpart of the recognition, etc. The very sensory appreciation of an object also requires movement, through the movements of the fingers, the eyes, the tongue, and (in animals) of the ears and nostrils. Tactile and visual appreciation, for instance, are impossible without the movement of the corresponding receptors. An image is no longer perceived after a few seconds, if it is maintained fixed on the same retinal points by preventing its displacement over the retina through the fine eye movements. In a limb paralyzed by severance of the motor nerve fibers, the tactile appreciation of an object fails if the movement of the object does not substitute for the impossible movement of the receptors on it. An early deprivation of eye movements definitely also impairs later the ability to appreciate the visual properties of an object, visual perception. The same occurs in the case of tactile sensation: early tactile deprivation induces with the passage of time an alteration of body and limb posture (thus producing abnormal muscle contractions and movements). Moreover, our internal organs participate in sensation (as they do in perception and recognition) not as a consequence of it, but as part of it. Changes of heart rate, of blood pressure, of respiratory rate, contractions of the bowels and of of the bladder, secretion of a great number of glands accompany to a greater or lesser extent all kinds of more or less defined sensations of uneasiness, of expectation, of anxiety. An object presented to our senses may not be perceived as such at the level of consciousness, and yet may be able to evoke undefined sensations, which are thus the ones making us aware of its presence.

Sensation, like motility, is subject to a type of control from the brain centers. (The term control is used here in a very loose and, probably, incorrect way, generally indicating that there might be influences of different types). This is based on centrifugal nerve pathways to sense organs. As the muscles and the joints feed back to the motor centers the information gathered by their sensors, so the sense organs receive signals back from the central structures, a function which still represents a physiological puzzle. Even if the physiological significance and the functional properties of these signals are not known, and the phenomenon is perhaps not as diffuse as previously thought, what is apparent is that there is some kind of interference with the sensory input to our brain. As a consequence, we are entitled to expect that our sensations evoked by one object will change in different situations, not only because our mood may be changed, according to the particular situation, but also because the sense organ itself is in a different receptivity state. Generally, the nervous structure associated with the sense organs is not complex, so that the feedback paths to these sensory organs might simply constitute the means of reducing their sensitivity. However, in the particular case of the visual organ, the nerve net of the retina might be compared to a rudimentary cerebral cortex, i. e. to a structure allowing a good degree of integration of the signals coming from the receptors with the ones of central origin.

### 3.1.1 Unity of Psyche

As for the psychic processes, their involvement is clear in the examples given above. The movements of limbs and body, as well as the sensations arising from different stimuli, acquire the right significance of purposeful, directed means of contact with the outside world, only through the ensemble of those central operations we call mental or psychic. There are many of the latter, at different levels. Some are also present in animals, but the great majority are peculiar to humans; some still closely connected with motion and sensation (although working through symbols), and directly representing external reality, others completely abstract creations of the brain. The same mode of proceeding by means of large, unitary involvement of the brain structures applies to the whole of the psychic function, which is now generally recognized to consist of innumerable and subtly differentiated nervous processes. The latter, however, are not independent entities, but different aspects of a unitary, complex function. Thus, memory association, abstraction, concept formation, and thinking, but also mental feeling, etc., are so many facets of the unique entity, our psychical life. It is only in teaching, in experimenting and in clinical practice that these processes are separately investigated (but, again, this is like separately examining the components of a sentence in semantics). An example of such unitary phenomena, still closely linked to physical reality, is the unconscious knowledge that man has of his own body as a unitary physical entity. In our brain we seem to keep a kind of scheme of our body, in which different parts constitute a whole (our "physical self") and which makes us act accordingly, but altogether unconsciously, not allowing freedom in either motion or sensation to any of the parts. Only in particular neuropsychiatric syndromes do we realize how such a body scheme may get lost, the patients acting as if some parts of the body were non-existent (for instance, unconsciously neglecting one half of the body or of the face). Implicit in that unitary feeling is the participation of conscious and unconscious sensory experiences and motions of the different parts of the body, through which each part makes "its own being there" a well-integrated component of the "whole being there", the whole of this physical self being represented by the whole of the brain. So, it seems that normally everything involved directly and indirectly in the physical relationship with the outside, is utilized for an abstraction, i. e. the concept of the unity, or singleness of the being, as distinct from the plurality of the outside world. Though abstraction is the basis of the many psychic processes typical of the human beings, this type of abstraction maintains, in the particular case, a very primitive character, since it is endowed with the most primitive and powerful species-preserving significance (and this might be the basis for finding a somewhat similar syndrome in primates as well, as a consequence of cortical lesions). This example fits into the discussion of the powers of the brute and the human "anima", extensively treated in the Appendix.

It is accepted fact that the psychic processes constitute a whole, as well as that even the most rational of them involve, to a greater or lesser extent, the participation of emotional processes. There is no instance in which we do not experience a feeling of pleasure or displeasure, of well-being or of uneasiness, of attraction, of "adherence" to a situation, to a talk, to a thought, in which we do not give our assent or dissent. This requires that the different parts of the brain are simultaneously involved. The human brain has the capacity for abstracting from particular situations, and thus for generalizing its responses to a large variety of symbols through which those situations are represented. Reading a story, or hearing a description of events, reading or hearing a poem, or music, may evoke from our brain the same reactions as if the facts were really unfolded under our eyes, with the "dramatis personae" moving and talking. The brain itself becomes a direct participant, with rational and irrational processes, with its feelings of pleasure or distaste, with its external signs of emotion, laughing or sobbing, causing the heartbeat to speed up or slow down, the respiration to pause or sigh in relief, being thrilled with interest, or yawning in boredom. Albeit temporally separated from the direct experience, the brain can also evoke by itself, symbolically, whole situations, recalling them from memory. What is peculiar to man, is that the brain creates in itself altogether new situations, again with the same attendance of feelings, emotions and internal events, as artists actually do, by which the very process of higher abstraction, the creation of ideas, reveals its nature of whole-brain concern. The most severe psychic syndromes are characterized by a "schisis" between these processes.

A direct consequence of the principle of "global" organization, as I have tried to outline it so far, is the failure to strictly localize a specific "seat" for each of the innumerable functions. At the beginning, neuroscientists were encouraged by successful attempt to produce paralysis and anesthesia through lesions in different places, mostly in the cerebral cortex. These results fostered a number of lines of research with similar aims, in the belief that to each of the specific functions corresponded specific and localized "centers". Piecemeal, all evidence now indicates that circumscribed lesions of neurons anywhere in the brain do not succeed in abolishing the control of movements, nor in attaining total suppression of sensation or of acquired memories. It is true that as consequence of lesions motility and sensation may be deeply affected or almost suppressed, so that one may talk of paralysis and of anesthesia, but this occurs most easily by interrupting nerve pathways (such as the efferent motor tracts, or the ascending sensory fibers), not by destroying localized brain areas. In other words, there are key points in the brain where lesions may be effective by interrupting the flow of information, just as the interruption of main roads may prevent a stream of trucks from supplying large sections of a town. The same concept of key points, however, does not seem to be applicable to psychic processes. There is a basic difficulty in investigating the latter, residing in our still extremely poor knowledge of their operational mechanisms, so that there is also a lot of arbitrariness in the tests of performance for these higher functions. Furthermore, it is exceptional to have the opportunity to locate lesions in man, since legally no harm can be caused to the human brain, except under severe and precise constraints (therapeutic and diagnostic procedures), while in animal brains we can only investigate something bearing an analogy with, but hardly corresponding to, the lower levels of the human psyche. Nonetheless, evidence from pathology and neurosurgery in man demonstrates that even vast destructions of brain tissue do not altogether abolish processes like memory, abstraction and thought. Large losses of neurons due to ageing (up to 30% of the brain mass), excision of parts or of an entire brain hemisphere, pathological degeneration, (even if associated with personality changes and severe dementia, and interfering with the access to consciousness) though deeply impairing communication between the brain of the patient and that of the examiner, still allow us to detect traces of those fundamental operations. (We shall resume later the problem of lesions and their significance in terms of brain function, particularly in as far as the "remaining capacities" are concerned.

## 3.2 A Whole Is More Than the Sum of Its Parts

Let's now delve into the deeper significance of "organization", which has so far been identified with the principle of unitary operation. This does not mean simply the simultaneous operation of the parts, because it would be equivalent to say that the whole of which we talk is nothing more than the ensemble of the structures, with all their connections. Identifying the function with the structure might be a philosophy, as we shall see in the next chapter, suitable for conducting some types of experiments, but would leave altogether in the shadow the most important properties which characterize unitary operation in a structure like our brain. The basic principle is that all the parts constituting a whole concur in the unitary operation in such a way that the result cannot be foreseen simply through knowledge of the individual operations. Clearly, in order for this to occur, the system must be constituted of differentiated parts, or properties; if it were not so, the result of the whole would be contained beforehand in the activity of the individuals, and could be represented simply by the linear sum of them. As a matter of fact, if the parts were of equal standing, they could also be interchanged, in principle if not in practice, there being no reason for anyone to be connected with any specific other, nor for a temporally ordered operation of one in relation to others: that is, there would be no organization. By analogy, for an engine made up of a large number of parts, each of them performing a different specific partial function, there is only one way of assembling the parts, the one corresponding to the "principle" of the engine, in order that the engine may function as a whole (or at all). Every other combination will be altogether ineffective, since none of the parts can be interchanged with others. Yet, in this principle there is still something more. Because of the uniqueness of assemblage, the system is clearly working on a non-linear basis (an engine will certainly show an all-or-none way of operation, even after the right combination of assemblage has been found, if the quality of the components is defective). We could compare our brain, however, to a system in which a defective component does not necessarily prevent the others from operating, thus operation is no longer on an all-or-none basis, albeit still being of a non-linear type because of the faulty components. This fact is hinting at some (perhaps a high) degree of redundancy in the number of the available components for each partial function. The main point is that while the term non-linear can safely be used in connection with every measurable parameter, since it actually means that the effect measured is not linearly proportional to the agent, when dealing with complex brain functions, where there is a multitude of agents simultaneously present, it seems more proper to talk of potentiation, of cooperative operation, of organization, terms which refer explicitly to the relationships between the parts of the whole. It is at the level of the organization that one jumps from the concept of "quantity" to that of "quality". Individual operations may be quantitatively enhanced (a positive power) or depressed (a negative power) by the activity of other individuals, but in the whole complex the different distribution of the quantitative effects, that is the different pattern of quantitative relationships between the parts (the parameter stating the conditions under which they are operating), finally results in qualitative differences, i. e. in different functions of the whole. Thus, the principle of "global" organization (of the whole), which would seem to divert our attention from the individual parts, actually states that the manifold functions of that whole can rest only on the suitably-varying functional relationships between the parts, and consequently that no function can be defined through the behavior of only one (or a few) of its components.

These conclusions are reinforced by other considerations. As said in the previous chapter, the thousand millions of neurons of the brain are always active, so that the overall impulses in the ensemble of nerve twigs can be assumed to be billions per second. Throughout the day, this activity changes quantitatively only little (as shown by the scant difference in the consumption of oxygen by the brain in different situations), although it becomes differently distributed among the various structures, strictly depending upon the situation. As a matter of fact, the various functions performed in 24 hours (feeding, sleeping, waking, social intercourse, etc.) require the neurons to be differently grouped, and differentially active for each function. Therefore, if anything must differentiate the activity of the brain at different moments, can only be the redistribution of the quantity of nerve impulses produced within the multitude of nerve circuits. The relationship between the parts of the system represents the fundamental concept, inasmuch as an understanding of the working principle, and comprehension of the "whole" operation, cannot be based on the amount of activity of this or that part of the brain, but only on the whole pattern of distribution of the activities. A picture remains essentially the same, whether printed in a darker or a brighter tone, and likewise a melody whether played loudly or softly, because what makes up the picture and the melody are the relationships (in space and in time, respectively) between their components. It is the same for the nervous system: its organization, through which it performs all its functions, provides for each function a definite plan of relationships between the subsystems; therefore, it is not the characteristics of the components, but the pattern of the relationships between them which represents the factor of relevance upon which we have to base our measures.

In the course of the last 40 years, patterns of activity in different brain subsystems have been detected, which correspond to what we call physiological states (e.g. arousal and different stages of sleep). These states, then, can now be defined (still, rather imprecisely) through the corresponding patterns of activity distribution (in Chap. 7 we shall deal at length with the definition of states). However, each of these states is also represented by an ensemble of "acts" specific to that state, which we call the "behavior", each of the acts being the result of the operations of the brain at that time, i. e. the whole behavior is the expression of the activity of the whole brain. As a consequence, what is again essential is the pattern of the relationships between its different components, through which we can compare individuals of the same and of different species. Let's take, for instance, a monkey. The difference between it and a human being is not that it sees a fruit differently, or that it moves its muscles differently, but that its general performance, as we can detect from its behavior, resulting from the interplay of all its structures, is different. It is the organization of the monkey's brain which "makes" the monkey. Yet, even if the monkey's brain is organized at a lower, or simpler, level than a human's, it is still a whole, complete in itself, and as such, as a unity, can be compared with another similar (monkey) and different (man) unity, provided we identify this unity with the pattern of relationships between its components.

## 3.3 Maturation and Learning

We are now confronted with the problem of how the organization of the brain happens to be the way it is. Let's start with the situation of the brain at birth. It is a fact that at birth the brain already has all its neurons, and no further generation of nerve cells takes place thereafter. This simple fact is remarkable, since certainly the human newborn has very little or nothing of the behavior of the adult, whereas in some species the newborn animal may exhibit only minor differences from the adult. We have two factors which can explain the differences between a newborn and an adult: structure and organization. The brain of the newborn is much smaller than that of the adult. As a consequence, there must be growth (which actually occurs), however not in terms of number of neurons, as previously said, rather in terms of the branches of dendrites and neurites. (Much is also growth of glial cells.) A histological picture shows progressive development of the branches which, at birth, are only beginning to sprout, at least for most neurons. This process is called structural maturation and takes a variable time to reach completion in different parts of the brain. Besides the structural aspects, that is, the formation of connections between the neurons, there is also activation of the newly formed connections, and of those already present at birth, by maturation of the chemical factors responsible for the transmission of the impulses at the synapses, i. e. chemical maturation.

The process of maturation of structure and function is best appreciated in those beings whose behavior matures soon after birth, that is in animals, where structural maturation merges early with another process initiated immediately after birth: learning. As demonstrated in an already classical example, by observing the behavior of newly hatched chickens, we can appreciate differences from an adult animal, e. g. in the ability to peck. These differences disappear in a matter of a few days, it being understood that the acquisition of adult behavior is partly due to structural (and probably also chemical) maturation, and partly to the practice of pecking during the first few days after birth, i. e. to learning. That these two processes exist there is no doubt, the point is to what extent each of them contributes to normal adult behavior. Since maturation of structure is a process which is already present before birth, and continues afterwards, whereas pecking can be initiated only after birth, it is obvious that practice (the more simple and basic method of learning) intervenes by shortening the maturation period. That this is actually the case is shown by the lengthening of the interval necessary for the appearance of adult behavior in those subjects which cannot practise (that is, learn) immediately after birth. From this point of view, learning is nothing more than a mechanism of potentiation, since it produces acceleration of an otherwise physiological process, organization, which must await the maturation of structure before it can proceed.

The same applies to the human being, although maturation and learning take much longer than in other beings. The main type of organized human behavior shown at birth consists of the signs of discomfort, which are certainly of high survival value, since through them the baby evokes the right behavior from the parents, i. e. feeding and caring for it, whereas left to itself it has no help from within, and is bound for certain death. Until a few months of age, the human infant is not even able to crawl, and needs to be carried around; it still needs to be fed and kept warm, and protected in various ways. Therefore, we are perfectly entitled to conclude that, in the human, maturation of structure, of synaptic mechanisms and, thereafter, of organization, is an extremely slow process. Since exposure to the necessary conditions for learning a behavior (including teaching by the parents) is continuous, we must ask ourselves whether practice is actually helping structural and chemical maturation. This is exactly what seems to happen. There have been many observations in humans in pathological conditions, (as well as a large number of observations in animals to back them), all pointing to the same conclusion: if for some time (months) a newborn is prevented from obtaining sensory experience, e. g. through the eyes, or from performing hand and eye movements, two types of permanent alterations result. First, in some of the neural structures involved, the neurons fail to develop their connections, and may eventually fall off; second, in the sensory and motor functions, the subject, although he may have a normal vision of the external world, may be unable to visually recognize objects and faces, and to suitably use his hands. This reminds one of the old statement of Aristotle, which is in Latin "Quibus deest unus sensus deficit una scientia" (that is, those for whom one sense is lacking are deprived of one science). I would rather interpret the sentence by extending the effects of sensory deprivation in a single organ, such as the eye, from the single science (vision) to the "unitary" science, that body of knowledge which is constructed day by day on the basis of what one learns every day, and to which vision contributes, integrated with other functions. This would fit in with the observed, although not easily detectable, deficits in the psychic domain, which need refined tests in order to be appreciated. In this context, one thing must be clear: in natural surroundings, learning any specific behavior is accomplished while we are exposed to a wide variety of conditions. Thus, the behavior does not depend upon only one trigger stimulus, but results from the interplay of a multitude of stimuli, integrated within large portions of the brain. In other words, each particular behavior is the outcome of the whole brain's activity, and if the latter has been affected by the absence of one sense, every behavior will be affected as well, to a greater or lesser degree.

However, the bulk of maturation of human nerve cells does not seem to take much longer than in brutes. The longer time necessary for the full development of functions in humans would seem to be, in effect, a learning time, a maturation not of structure but of organization, which can of course be completed only after the structure is mature. On teleological grounds, this should be perfectly understandable. The human offspring does not need only to walk upright and put food in his mouth, as monkeys do. Maturation, and therefore organization, must proceed far enough to permit him to live in much more complex social surroundings than monkeys, and to learn to construct symbols, language and perform abstract thinking.

Speech seems to be the strongest potentiating factor in the organization of the human brain after birth. Up to two or three years of age, human infants and adult apes may seem to perform at approximately the same level, although, of course, humans are always more proficient at communication and learning skills. Afterwards, however, the human's proficiency seems to shoot skywards, and the baby learns to do what a monkey will never be able to do. Having learned how to master symbols, the human brain seems to know no limits in its capacities. In a much more sophisticated way, the situation reminds one of the increased proficiency in learning simple behaviors of those animals which have previously been habituated to the "experimental set" (that is, to similar types of situations in which they are asked to learn to do something). However, while the adult monkey does not go much further than learning an experimental set, a four-year-old baby may even learn to read, to play the piano, or chess, and to do mathematics, and starts to think autonomously.

Perhaps, saying that the human brain might have no limits to its learning capacity might sound too committed. Let us try to approach this problem and see whether somewhere, or somehow, a limit can be assumed to exist. We have seen that learning by practising fosters maturation of structure, i. e. the formation of the necessary connections through growth of neuronal branches. Learning is, thus, fostering new patterns of organization, which result in behavior. It might be, then, that every time one learns a new behavior in adult life (when structural maturation should be well over), newly formed connections are necessary in order to support a new organization followed by that behavior. In other words, learning should always require an increase in structure. This increase, which after all would correspond to a "retarded" structural maturation, is unacceptable in terms of an increased number of neurons (with age the latter progressively fall off, thus exhibiting a tendency opposite to the growth of structure) and there is little evidence for further growth of branches. This conclusion should be modified, describing as "new" the development and the protrusion of synaptic "spines" from dendrites, and, more probably, the formation of preferential channels for the nerve impulses in the pre-existing neuronal net, based upon specific membrane receptors fostered by the presence of the corresponding chemical mediators, or of modulators. This hypothesis can account for the fact that it is always possible to learn new behavior at any time, even late in our life, through associations of conditions, no matter how strange and illogical, as is the case with conditioned reflexes. This property too is included in the term "plasticity", meaning that at any time organization (at a somewhat simpler level) may continue, if fosterd by suitable stimuli from the surroundings. That it might, after all, be a chemically-based phenomenon could be suggested by the rather quick disappearance of newly acquired behavior, when it has not been required by circumstances for some time (but certainly, the exact mechanism needs further investigation). There is another consideration. If it is assumed, as was said before, that behavior is the result of organization fostered not by a single trigger stimulus, but by a multiplicity of stimuli from the surroundings, then we should expect that every being will exhibit the correct behavior in every possible situation only if he has previously been exposed to the widest variety of stimuli, that is, if he has had the largest possible experience. This is actually what can be demonstrated experimentally on animals: populations free-bred in stimulating backgrounds exhibit greater capacities for learning and for responding to the surroundings with appropriate behavior, than their litter mates reared in isolation in dull surroundings.

With this, I do not want to convey the belief that behavior is determined only by interaction with the surroundings. There are certainly other sources of determination, as will be made clear in the Appendix. The brain, for its learning capacity, has been compared to a "self-organizing" system, i. e. to systems which reach their ultimate (highest) level of performance by interacting with the surroundings, or in other words, systems which organize their knowledge and their behavior by themselves with what the surroundings offer them. However, in the self-organization thus conceived is not implied to any extent the possibility of choice. On the contrary, the system is thoroughly dependent upon the surroundings. This point seems to mark the difference between the brain and man-made, non-living, self-organizing systems. The brain fares very poorly with respect to the constancy of its interactions with the outside world for a number of reasons; particularly, a poor memory, poor precision of its sensors, the presence of interference from many sources in its decision-making circuits, the presence of factors other than chance and weighing heavily against probability. We do not know at present which are the causes and which are the effects. The point is that the brain has developed (as an evolutional feature?) the capacity for selecting from the surroundings the stimuli to which it reacts, neglecting a host of others. In this sense, the brain is to a far greater extent a "self" organizer than any man-made device. Merely for its large degree of independence from the external world, we should refer to it as an autonomous organizer. The world teaches and the brain learns, but in learning it orders material in a way that cannot be described as simply "registering", and its operations and results are not determined, except to a minor extent, by the simple presence of the external stimuli. (See also the Appendix.)

## 3.4 "Human Whole" and "Brute Whole"

The fact that our considerations have led us to conclude that the brain normally operates as a whole, does not mean that the functions of which the brain is capable and which it is continuously accomplishing, cannot be arranged in order of complexity, from the lower, simpler acts (the reflexes) to the higher, and certainly more complex ones involving the formation of ideas and thinking. However, if we maintain the concept that neuronal operations are a whole-brain concern, we cannot also presume that a rigid distinction between the functions can be maintained, even in the case of the most "distant" ones. Some of the brain functions are represented by very simple and basic acts of life, like feeding, mating, life-preserving reactions, etc., while others belong tho the "higher spheres" of the human intellect; the first ones being in common with brutes, the latter being exclusively man's. The fact that we share some functions with other mammalian and non-mammalian species should not lead us astray. We said at the beginning of the chapter, that it would be foolish to try to peel off the newer brain structures with the aim of reducing the human brain to the level of a brute's, since old and new have grown in the same spaces, not simply adjoined, and have developed a thick maze of interconnections. However, apart from the practical impossibility on anatomical grounds, it would be utterly impossible on functional grounds as well. The way the human brain operates in performing those "lower" acts, cannot be the same as the brain of a frog, a sheep, or a monkey, because these functions, in man, are now subject to a variety of internal conditions, created by the presence of the newer and more complex structures peculiar to him. These structures do not simply add their ability to construct ideas or abstractions to the act, e. g. of feeding. Rather, the simple feeding is performed by structures which, through their fusion with newer ones, have in some way acquired some of the latter's properties. There are, so to speak, "ideas" in the feeding (or traces of them), which make of the act something typical of man and less similar to that of a brute. Likewise, the reverse must be necessarily true also, since in every intellectual operation of ours there always will be a trace of our brute's being, a trace of our "lower" habits. Thus, we must conclude that the brain of the intellectually powered man performs feeding, senses, and controls the movements of the muscles, etc. in a more complete and, we could also say, more "perfect" way than birds, dogs, or monkeys can ever do. These concepts stem coherently from the way the structure of our brain has been formed, but, as concepts, are not new: they were expressed by the philosophers of the Middle Ages, who had already considered the development of the powers of the human mind as the last step in a continuum of increasingly "powered" brains of non-human beings. (See also the Appendix.)

A short, final comment on a problem which should not belong to this chapter, since it is deeply entrenched in the philosophical aspects of our cerebral activity;

that is, how man utilizes the organization of his nervous system. This question has two possible answers, which are not necessarily mutually exclusive, and still nowadays correspond in their essence to the two main, old views about man's stand in this world:

i) The brain's organization allows only the mechanistic development of nervous events, evoked by suitable stimuli or (better) by suitable constellations (or complexions) of internal and external stimuli, i. e. by a "situation". This scheme can probably also account for some degree of unpredictability, because of the variability of the complexions, and the extreme complexity of the central nervous structures and their organization.

ii) A "free will" can control the chain of events. This statement could simply mean that any mechanistically determined chain of events within the brain can be interrupted and redirected by a voluntary act.

Of course, it is at the level of this term, "voluntary", that the opinions diverge. By that is meant decision, unpredictable and autonomous, therefore not determined by the complexion of conditions (the stimuli). However, it could always be maintained that the exact complexion (particularly of the internal conditions of the brain) cannot be known with sufficient precision, but that if it were known, the decision would be found to be completely determined by it. This interpretation would again allow only the purely mechanistic view of our brain's function, one that considers man to be a being fully dependent, in the end, upon the environment, since the internal conditions would also be primarily shaped through interaction with the surroundings, "printed" early (or genetically inherited) in the brain, or taught later in life. So, there would be no true freedom in the so-called voluntary act. By what specific neural mechanisms a free-will decision could be originated, I do not dare to venture. It is far beyond the scope of this monograph to further pursue this subject, which still constitutes a much debated question among philosophers and theologians (besides, of course, neurophysiologists). Only in the Appendix, will points of view that are not strictly neurological be briefly discussed, in line with the general theory I shall go on expounding in the following chapters.

# 4. Short Review of Approaches and Methods

## 4.1 The Standpoint "Structure = Function"

I said at the beginning, that a science is progressing when it produces hypotheses that can be put to the test and, thus, proved or disproved. However, in order to make hypotheses one must have a starting point, a general theory, a philosophy of research. The neurosciences, though an applied branch, have not developed independent techniques as such; the absence of a general theory or frame has certainly favoured the development of largely differentiated experimental approaches (which is by no means a disadvantage, because multiplicity of approaches is always desirable) but has also induced a tendency to exploit the advancements of other fields, such as physics, chemistry, mathematics, computing etc., in a completely independent way each of the other. There is also an intrinsic danger in this situation, i. e. the neuroscientist does not easily perceive that with anyone of the widely differentiated methods, he also implicitly assumes a whole package of concepts, and particularly of limitations, inherent in the method itself.

I shall try in this chapter to review the main lines of methodological approach to brain problems, beginning with the oldest one, classical anatomical research. Anatomy did not progress much until it was unmistakenly demonstrated that the nervous system is made up of neurons, i. e. individual cells, connected to each other however, through a terribly intricate maze of extremely fine expansions of the cell bodies. At a time when the modern electrical methods of recording the activity of neurons were still unknown (the instruments being represented only by capillary electrometers and slow galvanometers), the nature of the electrical signs of nervous activity could not be investigated or understood, and the study of anatomical structure dominated the whole of neurological research (with the possible exception of psychology, just at its dawn). It is no wonder, then, that the neurophysiologist tackled the neurological puzzle by trying to pin functions to different structures. Thus, he started the practice of placing lesions, rather grossly at first, in those structures which were externally visible (and which were soon demonstrated to be highly internally heterogeneous) and later on, with the progress of microscopy, in increasingly better localized centers.

On the other hand, earlier clinical observations somehow reinforced the belief that the brain consisted of an ensemble of centers for the individual functions, and, even if not clearly expounded, this might be taken as the first and (so far) sole theory of the brain functioning. The fallacy of the theory was demonstrated little by little, when working hypotheses which tacitly assumed it failed to precisely localize functions in anatomically restricted structures. As I said in the preceding chapter, every function is not uniquely performed in one confined center. It results, rather, from the cooperation of different structures, sometimes widely dispersed in the brain mass, each contributing with its particular properties to the multifarious aspects of a function (e. g. motion or sensation). The slow realization of this did not eliminate the method of localized brain lesions, because it is assumed that, even if complete disappearance of a function is no more to be expected, at least a "deficit" in its execution should be detected, due to disappearance of the properties of the structure destroyed. Therefore, producing localized lesions is still a favorite and much cherished method. Its usefulness is probably unmatched in experimental psychology, the more useful the smaller the lesions, though requiring a continuous refinement of the tests for detecting functional alterations in the animals, with whom communication by language is altogether impossible, and in whom the objective signs might be extremely scant. Probably, different methods of appraisal of results will also be required as well, since at present they are all based on statistics.

As for the significance of a "deficit" after a localized lesion, it must be clear that the absence of some particular property after the destruction of a small part of a structure, does not allow us to conclude that the property was actually represented in, or dependent upon, that structure, in other words, that the structure was the center, or the "site", of that property. Things might not work that way or be as simple as that. For example: if a lesion in some rhinencephalic structure is followed by increased tameness, or by increased fierceness, it does not mean that the corresponding centers have been destroyed, or that tameness and fierceness depend solely upon those structures. By analogy, let us consider a slit in a sheet of rubber kept under some tension. If the gap thus opened does not widen and the sheet does not tear, one might even think that little or no harm has been done. However, if suitable methods are available, one can detect that the stresses and the strains imposed are differently distributed. This is, actually, a "new" pattern of stresses and strains, and it would not make much sense to maintain that it is a "deficient" pattern of some kind. Conversely, if the sheet tears, this is not evidence that the damaged area was sustaining all the stresses of the applied force. In other words, after a brain lesion, it is probably the change in the function of the remaining structure that counts, rather than what has been lost: the result would be not a defective, but a "different" function.

Nowadays, the equation "structure = function" no longer refers to individual centers, but to the (more complex) circuitry of the neuronal connections. The more recent technical developments of joint neuroanatomical and neurophysiological research, have succeeded in tracing the finer connections between neurons, an impossible feat with the methods usually employed in tissue staining. The beautiful methods of earlier anatomists, Golgi and Ramón y Cajal, based on metal impregnation of the tissue, although giving detailed information on the course of the fine neurite and dendrite branches, are completely devoid of any functional reference and do not allow quantitative conclusions. The new methods, based on histochemical reactions of molecules absorbed by the cell bodies or the neurites and migrating along the fibers, have now become standard practice, because they allow one to detect: i) direct connections even between structures very much far away from each other, without the need to demonstrate the thin nerve filaments connecting them; ii) chains of functionally connected neurons. To the same purposes are used radioactive chemicals absorbed by the tissue according to its functional state, showing their distribution through autoradiographic techniques. The intrinsic difficulties met by anatomists and physiologists in their incessant joint task of tracing the nerve connections, has brought with it an increased conviction that the anatomical and physiological problems are inextricably entwined with each other, and that knowledge of the structure is a sufficient requisite for knowing the function as well.

## 4.2 Electrophysiological Methods

In the last four decades the far more useful and widespread method of investigation has been proven to be the recording of the electrical activity of neurons (electrical currents and potential changes in the cells and in the nerve fibers). This approach quickly gained the favour of neuroscientists as soon as the technical revolution in the electrical recording instrumentation introduced by the thermoionic tube had brought the possibility of detecting voltages in the microvolt range and in a wide spectrum of frequencies. Thus, direct recording through suitably constructed electrodes was made possible, both from large populations of neurons as well as from single cells. However, it was not until the end of World War II that electroencephalography, on one side, and single neuron physiology, on the other, began to build a systematic clinical and experimental neurophysiology, and entered into a true boom era.

The nature of the method, recording from the surface of the brain or from the scalp, renders electroencephalography best suited for human investigations. It was from the very beginning more and more concerned with deciphering the significance of the brain "waves", the electroencephalogram (EEG). Its experimental counterpart, the single neuron recording, revealed an unmatched usefulness for exploring neuronal activity in deeper parts of the encephalon. These two lines, the clinical and the experimental, have since further differentiated, each following its own path of development, by refining the techniques of recording and of analysis of the data, but with different journals and languages, so that they now almost constitute two different sciences. One deals with signs of brain activity which seem to have acquired more the value of symbols, far away as they are from their source and from the real appearance of the phenomena at the origin. The other deals with the more direct representatives of neuronal activity, the nerve impulses, following them through the maze of connections, and building what we shall call from now on "analytical" neurophysiology. We shall treat the two differentiated lines of investigation separately, beginning with the experimental approach.

### 4.2.1 Analytical Neurophysiology

Although the electrophysiological methods did not have the specific purpose of tracing the finer neuronal connections, one of their most outstanding achievements so far has been to define the scheme of a large number of circuits, which subserve specialized motor, sensory and integrated functions. The main reason is that, there being direct and indirect connections, it becomes necessary to know which is which, since the activity evoked in one neuron by the activity of another has, obviously, an informational content dependent upon the path followed by the impulses. The blooming of the tracer methods in neurohistology in connection with electrophysiological research, meets the urgent need of defining the plan of the communication lines, in the hope of being able to read the information contained in them. Moreover, since the brain is an organ in which function is strictly bound to structural organization (but I always maintain it does not actually consist of structural organization alone), there is a widespread belief that a description of structure may lead without much further effort to the description of its function. That structure and function are strictly interdependent is surely a sound concept, realized in all organs and tissues, so we should not have any hesitation in accepting the same proposition for the brain. For instance, the distribution and the direction of the osseous lamellae inside a bone are a direct indication of its function, because they immediately reveal the stresses inside the apparently rigid structure, and the related lines of deformation. They indicate, moreover, what plastic changes are required when the equilibrium of the forces becomes altered. Pointing this out as a further analogy (see Sect. 4.1) with a physical system in which the distribution of forces is one of the main functional characteristics, has been done purposefully (and it will be clearer the more we proceed), since I have already hinted at the role I am attributing to the forces in the nervous system (Sect. 2.6).

Reading the information contained in a nerve message will not be an easy or, perhaps, a possible task. We have seen in the second chapter that the structure of the nerve signals is generally very complex, because of the convergence on the neuron's membrane of a large number of nerve endings from different neurons. The signal thus received is made up of a number of time patterns of impulses at the input, integrated all over the receiving cell membrane during some "characteristic" time (in the vast majority of cases an utterly unknown variable), to which corresponds a simpler time patterning of impulses in the output of the neuron. The difficulty is circumvented by restricting the task, i. e. by assessing only the type of stimuli to which the neurons are responding with a change in their ongoing activity, the presence of the response generally being determined through statistical analysis of the impulse discharges. In this type of research there are only two variables, the stimulus and the change of activity (the response). However, these variables are not always fully controllable, so that the correlation between stimulus and change, which is sought as a demonstration of the response, has in itself an unavoidable and variable degree of uncertainty. Furthermore, it is impossible to set up a suitable stimulus with sufficiently determined parameters in every neuronal circuit, this can, in general, only be done for the sensory system, by manipulating the stimulus at the level of the sensory organs' receptors. In the latter instance, however, it is not necessary to bother any more about the exact structure of the signal issued by the sense organ – since that signal will contain in any case the information supplied by the stimulus parameters – or about the structure of the signal issued by the central neuron as a response. As a matter of fact, it can always be assessed whether a neuron is responding to that type of stimulus, and if not (by changing the stimulus parameters), to what other types of stimuli (in other words, to what different information content of the input signals) the neuron is receptive. A typical example is represented by the investigations in the visual system, one of the favorite "playgrounds" of neurophysiologists. The transmission of visual information can be followed from the retina to the first subcortical structure, then to the primary visual cortex and other cortical and subcortical areas involved in the different aspects of visual recognition. Thus, according to what we were saying about signal's structure, at each neuronal station the incoming message is transformed in some way, the number of impulses at the output, as well as their frequency and time distribution, being different from those at the input. A retinal message carrying information about the intensity of a point of light illuminating small groups of receptors is not relayed as such through the first synaptic station, the lateral geniculate nucleus (that would make little sense), nor is it simply scaled up or down (that would also make little sense). Rather, it is integrated with the signals from other receptor groups, (which too would make little sense, were it not for the hypothesis, a sound one even if not clearly proved, that this might serve to increase the contrast between the strongly and the weakly illuminated retinal receptors). What's more, after the first station, the message content, the intensity of the light, seems to get lost. At the first cortical station, the information about intensity is apparently diminished, and the messages seem to be preferentially related to the spatial orientation of lights and shadows, through a further integration of the signals from different neurons of the preceding neural station. What one usually concludes is that the cortical neurons are "responding" only to stimuli with definite spatial orientation and dimension across their respective receptive fields. This is actually not an intrinsic property of the cortical neurons: simply, they must respond only to those types of stimuli, because the multiple inputs are organized so that the impulse firing follows only that pattern of synaptic excitation. The consequence is that the information content of a message, we should say its "value" for any given neuron, is not contained in the simple sequence of the impulses in an individual nerve fiber alone, but is determined by the space and time integration of the individual signals over the neuron's membrane.

The example of the visual system is a good index of the real value of analytical research. While tracing out the successive stations in the visual pathways, it also uncovers the different connection of each anatomical structure, to which it refers the differences of "function". However, the ways and the "whys" of these progressive transformations of the impulse patterns (we should say qualitative steps) are not yet clear, partly because we still ignore the actual synaptic conditions which shape the nerve signals in the output neurite, but principally because the primary logic of the system is unknown, and cannot possibly be revealed by analytical methodology alone. As a consequence, the analytical approach has been producing a wealth of results, as a method of fine mapping of connections and as an analyser (justifying its name) of the properties of individual neurons, but is unsuitable for deciphering the basic principles of nervous function.

Working with the analytical method, the conviction that structure and function are nothing but two interchangeable aspects of the same reality is certainly reinforced by the results, and it is easily taken as a general principle itself, according to which all there is to know in the central nervous system is the precise map of all the connections involved in each function. This would lead us to assume that what differentiates the functions would be simply that each one involves different sets of structures. We shall see later on how this statement ought to be modified. By putting all the weight on the anatomical connections, one does not sufficiently take into account that the synapses are also organized in chemical systems, nor that the activities of different brain parts are influenced, in a still unknown way, by the modulators released under the influence of unknown factors. Furthermore, the analytical approach, though certainly basing much of its penetrating power on the knowledge of the structure, has produced evidence of phenomena which are not to be accounted for only through structure, such as the potentiating effects, the cooperation between structures, and the quality changes of messages. The method, therefore, should not be exploited as a tool for dissecting only structure, but also organization. However, there its limit also resides. The operation of dissection becomes cumbersome when applied to complex circuits, which must be analyzed in a sequence of steps without an overall view of them, and when dealing with nervous activities which are not strictly dependent upon a single event (such as a known stimulus of some sort), but result from the interaction of different circuits in which an "induced" change cannot appear as a plain and time-concentrated response. It is extremely difficult to analyze the latter situation, because there is never a clear definition of the nature of the hidden induced activity, of its composition, or of the ways it was built up, within a limitless, undefined ongoing activity. Finally, analytical investigation applied to neural activity can only be quantitative, but the statistical approach that one generally has to use for the quantitative appraisal of the data is, very much probably, inadequate for the purpose.

### 4.2.2 Microelectrodes

The analytical method has its principal tool in microelectrode recordings of unitary activity (the impulse discharge of individual neurons). Apart from some trivial, technical difficulties, a main problem with the use of microelectrodes is represented by the need to know exactly which neuron is being monitored. The recent progress in histochemical tracer techniques seems to have given an answer, provided the tracers are injected through the same microelectrode (a fine micropipette) which is used for recording.

At the basis of microelectrode recordings is a more or less tacit assumption that nerve impulses are the sole, or at least the more representative, elements of the neuron's activity. The nerve impulse actually represents the output of a neuron, as it results from the membrane processes initiated by excitatory and inhibitory synaptic activations of different origins (made visible by the excitatory and inhibitory postsynaptic potentials). Thus, the nerve impulse constitutes simply the message sent from one neuron to another and cannot represent the whole neuronal activity. In view of the fact that organization (essentially based upon the reciprocal relations between neurons) is the main characteristic property of our brain, the messages functionally linking one neuron to others should actually represent the essentials of its activity. However, this conclusion would deprive the slower membrane potentials and the everchanging potential fields around the neurons of functional significance. This problem will be resumed later (Sect. 4.4).

The microelectrode approach also assumes, more or less tacitly, that each neuron is representative of a "class" of neurons having similar properties (generally, the way they respond to stimuli, their discharge rate at rest, or other functional characteristics), so that the microelectrode is actually used for defining classes of neurons. Of course, these classes are somewhat arbitrary, since the criteria for class identification are determined by the experimenter, and it can be assumed that in some cases they might even be irrelevant for the real function of those neurons. It may occur, also, that classes do exhibit consistency within some well-defined experimental conditions, which vanishes as soon as the situation is changed. To this phenomenon is certainly related part of the observations that the same neuron, when tested on different occasions with the same type of stimuli, does not necessarily exhibit the same behavior. There are changes of the latter occurring during the same experiment, generally attributed to changes of the brain state (which can only occasionally be demonstrated) or of the local oxygen supply (which only sometimes could be ascertained).

However, it seems also reasonable to cast some doubts on the constancy of the behavioral properties of a neuron in every possible situation. We have only to think that the response of a neuron is a function of its input, which is itself a variable controlled by the experimenter to only a very limited extent, i. e. strictly limited to the stimulus parameters, thus not including the messages arriving from other origins. All these factors are likely to produce changes in the structure of the nerve signals at the output, so that even if in different situations a response of some kind can be detected with the usual statistical methods, its "significance" could be altogether changed, and unknown.

## 4.3 System Analysis

Analytical research has generated a particular type of investigation, perhaps the only one so far following a basic philosophy, the so-called "system analysis". It is, as the best offspring of the analytical method, a way of dissecting a complex system of organized parts, that is, a way of taking apart the components, of separating the individual functions, and of detecting how the functions are performed through the interconnections of the parts. It would seem that this is exactly what every analytical investigation should be aiming at and strive to achieve. System analysis in neurophysiology is backed by a large body of knowledge accumulated in the last few decades in the fields of cybernetics, and of mathematics of stochastic processes, which makes the method a powerful and penetrating tool. Basically, a complex system is considered to be made up of "black boxes" (structures with unknown content) and one aims at assessing the quantitative relationships between the output and the input of the box, starting with the simple assumption that the output should always be some function of the input. The immediate result of the method is the "transfer function" of the black box, i. e. the mathematical form of the operator which transfers the properties of the input to the output. The transfer function need not to be simple. Actually, it may assume a mathematically complex form. While, as stated, the method should give rise, in principle, to fully determinative relations, (at least there is no reason why it should not), in practice, however, it does not work that simply. Like every other physical system, the nervous system can never be so precise in its functioning as to exclude some degree of (seeming) randomness, and, as I have repeatedly stated, there is no way of knowing exactly the structure and the dimension of the signals at the input, apart from a few exceptions. Therefore, the relationships between output and input are better represented statistically, treating the variables as a stochastic series of events originated by stochastic processes.

The degree of that randomness within the central nervous system is enormously increased by uncertainties about the measures of the variables, input and ouput. So far, this constitutes the main obstacle to a generalized introduction of system analysis into the study of brain function. The method can, however, be successfully applied whenever the input and output variables are known with some precision, that is, in the sensory system at the receptor level, and in the motor system, again in the more peripheral structures. Until now, the motor system has provided the best results. The movement of a limb, or simply of a joint, is produced by one or a limited number of muscles. The motor nerves carry messages to the muscles, containing the necessary information for performing the movement with well-determined velocity, extent of displacement, and force to be applied. The sensory nerves from the muscle and joint receptors carry all the information about velocity of movement, displacement, and force at any instant while the movement is being executed. The sensory information is partly fed to the motorneurons, partly to higher centers, which, in turn, send back messages to the final motor structures in order that the movement may reach its final goal as programmed. It is nothing more, in this sketchy outline, than a servomechanism, though, as we have seen previously, in these automatic processes there is also a place for influences which sometimes completely deprive the mechanism of its automatic character. As a servomechanism, it lends itself beautifully to system analysis. Piecemeal, the transfer functions of different black boxes are being computed and the peripheral motor control will be sufficiently known for all practical purposes.

In the sensory system, things cannot work with equal precision. As we have seen previously, the quality of the messages seems to change when crossing the synapses central to the sensors, so that a purely quantitative relationship between output and input, while always possible mathematically, would not be as easy to correctly interpret. It seems that rather than transfer functions we need "transformation functions", i. e. operators which, in performing the transformation of one quantity into another, may simultaneously perform the transformation of the "quality", in order that the original information does not get lost while it is being combined with other information of a different origin. It is apparent that the main hindrance to the extensive application of the method is the need to know precisely the input message. Moreover, as in the motor system, the nerve message at the input contains information with a number of parameters, like instantaneous and peak frequencies, time to peak frequency, plateau frequency, dacay time, number of impulses per train, number of trains, etc. It is highly unlikely that all these parameters are transferred from input to output in such a simple way, that a single transfer function may take care of all of them. It seems much more likely that another concept (and method) must be applied, i. e. that of the transformation just mentioned, which, as we shall see in the following chapters, is based on somewhat different mathematics and concepts.

## 4.4 Brain Waves

The electrical activity of the neurons is not exclusively represented by the nerve impulses: the latter are generated in the neuron as a consequence of a predominance, at any moment, of the excitatory over the inhibitory synaptic effects produced by all the impulses at the input, regardless of origin. The biophysics of excitable membranes has led us to understand the fundamentals of the continuous fluctuations of the membrane's electric potential, which are predictable as to their time course when a nerve impulse is generated, but are not predictable otherwise. Thus, while the nerve impulse appears to be a fully determined phenomenon, the residual activity during most of the neuron's life would not seem to be subject to precise laws. Microelectrodes inside the neurons and in the immediate surroundings have demonstrated the potential field distribution and the ionic currents crossing the membrane, actually linking the inner to the outer cell medium and rendering the neuron, which still maintains its anatomical identity and its independence, part of a continuum. However, the situation is completely different when recording the activity of large populations of neurons, either directly from the cortical surface, or, more frequently, from the scalp with macro (large-surface) electrodes. The meninges, the cerebrospinal fluid, the skull, and the scalp constitute pathways for the short-circuiting of electric events generated in the brain mass, before they can reach the electrodes (generally more than 1 cm away from the source of the electric disturbances). The presence of different tissues has profound consequences on the recording of the two main types of event, the nerve impulses (developing a large potential difference across individual membranes, more than 0.1 V), and the much smaller fluctuations, (of a few mV) of the mebrane potential. The latter is an integrated phenomenon, the postsynaptic summation of the excitatory and inhibitory impulses impinging on the neurons, and of the corresponding fluctuations of the internal and external potential fields when the cell is not discharging impulses. The two types of events have different time courses: of the order of 1 ms, the nerve impulse (corresponding to a "frequency" of around 1 kcps); of the order of 20–2000 ms the slower fluctuations, the brain "waves" (corresponding to frequencies between 50 and 0.5 cps). What the electrodes on the surface of the structure (brain or scalp) are actually recording is not determined so much by the voltage amplitude, as by the frequency characteristics of the voltage fluctuations, because the attenuation factor of the voltage from the source to the recording point depends upon the velocity of the voltage change, i. e. upon the equivalent frequency. The attenuation factor being higher for the higher frequencies, what the surface electrodes pick up are principally the lower frequency events, those having smaller voltages at the source, but subjected to proportionally smaller attenuation through the complex of the volume conductors than the higher voltage, higher frequency components of the neural activity. Since the lower frequency voltage fluctuations are also attenuated, the resulting amplitudes at the electrodes are in the μV range (1–50). Consequently, these types of recordings may be polluted by electric potentials other than nervous ones (such as from muscles, dipole ocular movements etc.), and particular care is required for their elimination. There are also brain potential oscillations of frequency lower than 0.5 cps (the so-called "slow potentials") as well as true steady potential differences (which are different in different situations). The nature and the physiological significance of these slower potentials is not yet clear enough, therefore we shall not be concerned further with them in this short account.

Recording simultaneously with microelectrodes and surface electrodes from the same structure, reveals that the impulse discharge and the wave potentials do not maintain precise phase relations with each other, and sometimes none at all. This fact would seem to disprove the assumption that surface recording results from the summation of the graded postsynaptic potentials created by the input impulses, preceding the subsequent impulse discharge by the same cell. Simple consideration, however, actually shows that the surface recording has all the information related to the impulse activity, though not in a simple form. The postsynaptic potentials are graded because they are summable over their elementary components (within limits, also linearly), therefore they are related to the input impulses to the neuron, while they are never in simple relation to the output impulses, because the impulse discharge is always a non-linear function of the input. Thus, returning to the concept of the black box, and considering input and output in terms of impulse discharges, the postsynaptic potentials should represent the membrane process more akin to the transfer function of the neuron. The lack of synchrony between the activities of different neurons, even within a restricted volume, and the differences, at least quantitative, between the different transfer functions, largely account for the uncertain, or unsteady, phase relationships between the recorded brain waves and the impulse discharges. Moreover, each neuron being part and parcel of a continuous medium, its receptivity to the arriving impulses is also dependent upon the state of the continuum that it helps to determine, as well as upon the chemical relationships with the glia, and the action of the as yet unknown modulators on its membrane and its metabolism.

The range of uncertainty can be reduced to practically nothing in well defined conditions, principally repetitive sensory stimulation, slow-wave sleep and epilepsy. A sensory stimulus evokes in the central sensory structures (and in other structures connected with these as well) a response consisting of a series of potential waves with rather unsteady waveform. In correspondence to the latter, as usual, the impulses issued by the neurons exhibit no steady phase relations with individual components of the complex. However, with repetition of the stimuli one reaches a better stability of the waveform of the complex and, simultaneously, a better and steadier phase relation between waves and issued impulses (although different cells will exhibit different phase relations to the same waves). The reason is that by repetitive stimulation the impulses arriving at the synapses can summate at regular intervals through both the direct and the indirect pathways, thus eliminating much of the apparent randomness of the membrane processes following one single stimulus. To the reduction of uncertainty, contribute also, probably, the strong, homogeneous electrical fields created in the tissue under these conditions (see also the end of Sect. 2.6). At the time of arrival of the impulses, therefore of the occurrence of the neuronal response, a quantitative data analysis allows interpretation of the waveforms and the delay time of each component in terms of impulse discharges. In this way, one has succeeded in attributing the origin of the potential waves recorded on the surface to definite neuronal groups of the cortical layers, and in some cases also to subcortical structures, with a fair degree of accuracy and reliability, and with good confidence about the sequence of the issued impulses. This "evoked response" method (see further, Sect. 4.5.2.3) besides representing by itself an efficient test of the physiological state of the cortex, has undoubted value for the clinician, who finds in it the only source of information as to the distribution of the excitatory and inhibitory influences within the cortex. The simple electroencephalographic recording (EEG) alone does not allow unequivocal identification of the brain waves with the corresponding synaptic effects, because these are present both in the depths as well as on the surface, and the different orientation of the cortical gyri gives to cortical cell layers no uniform geometrical referability to the electrodes.

As anticipated, there are other situations in which the relations between brain waves and impulse discharges become more strict and predictable. They are not inherently artificial (as is generally the case with repetitive sensory stimulation). One is clearly physiological, i. e. sleep, while the other is clearly pathological, i. e. epilepsy (although in its mechanism we may see an exaggeration of the physiological way of reacting to stimuli). What unifies these two apparently different orders of phenomena are several features they have in common. The first one is that they consist of potential oscillations of high voltage and of lower frequency components (in comparison with the nerve impulses), mostly grouped in trains, with definite waveform and considerable stability, often with monotonous repetition (particularly in epilepsy). These characteristics represent what we call "synchronization" (which will be more extensively discussed in the next section). The term means that at the time of the large waves the impulse discharges are grouped predictably in some fixed phase relation to the waves, each neuron with its own. In the case of epilepsy, the repetition rate of the impulses within each group may reach the highest values possible, being much lower in sleep. A common feature is that, although occurring spontaneously, both sleep and epilepsy can be induced by low-rate repetitive sensory, or artificial, stimulation, a fact which leads us to consider the possibility of common, basic mechanisms for the stabilization of the waveforms in the evoked response process, and the appearance of large, stable and repetitive waveforms in sleep and in epilepsy.

## 4.5 Methods of Quantification of Nervous Activity

Every investigation in neuroscience proceeds by contriving to establish a correlation between two or more parameters, e. g. the activity of different structures, or between these and some other internal or external conditions. It may be that we are simply looking for a correlation between their respective occurrences, but more often we are interested in the correlation between their "degree" or "intensity". In this way, we may assess the effects of lesions, the consequences of stimuli, the development of activity in different parts of the brain, the progress of conditioning and learning, etc. In every case, establishing a correlation demands a means of quantifying the phenomena and the parameters. In some cases, one has to deal only with the number of events, so that one simply compares counts, but modern neurosciences more frequently deal with nervous activity in terms of electrical potential oscillations, as we have seen in the preceding section. Since the recordings are effected either through microelectrodes or large surface electrodes, a distinction must first be made between the discrete time functions, such as the nerve impulses, and the continuous ones, such as the low frequency, surface EEG recordings.

### 4.5.1 Quantification of Nerve Impulses

The quantification of activity in terms of nerve impulses does not present any particular problem, since by the nature of the process operating on them, they can be treated as units characterized only by their time sequence, that is, by their number and by the time intervals between them. The methods of quantification are based almost exclusively on a statistical appraisal of the impulse sequences, and of the statistical significance of the differences between such discharge trains. The real point, however, is the physiological significance of every manipulation of the data, i. e. the neurological counterpart of the statistical treatment. As we have seen, in the case of impulses issued from sense organs, the quantification of the impulse trains is a measure of the response of a cell to a stimulus with defined parameters (peak and steady state intensity, time to onset and offset, duration, frequency or repetition rate, etc.). But, in the central pathways, the characteristics of the stimulus as translated in an impulse train, might be overshadowed by the changing significance of the message while traversing successive synapses, and the significance of the new message in physiological terms is not known in the vast majority of cases. Thus, the quantification of messages can be used for determining whether they are different, but does not help much towards knowledge of their significance and mechanisms. Furthermore, as stated before, the recording of single cell activity has the basic significance of building classes of neurons with homogeneous behavior. The difference between these classes is not only a qualitative one, as when the neurons respond to different types of stimuli, but may well be a quantitative one: for instance, when the neurons respond to messages differing as to the range of the intensity values they are relaying. If the situation is not known beforehand, i. e. if it is not known in advance the type of nervous message to which the statistics are applied, the results could be altogether false. In the first instance – classes differing in the quality of the stimulus- the data from cells of one class can be pooled without danger. In the second instance, on the contrary, the very subdivision into classes becomes a really arbitrary decision, because, first, it is necessary to define the allowed fluctuations around an unknown average of the messages relaying ranges of intensity.

### 4.5.2 Quantification of Continuous Time Functions

With the continuous time functions, such as the EEG, one is on quite different grounds. The data are not representable as a number of elementary events. Rather, the voltage waves constituting the EEG represent a summation of unknown amounts of membrane potential fluctuations in all the cells of the "territory" covered by the electrodes (the true extension of which can be determined, although not easily and not always, only through indirect methods). There is an old assumption – the first interpretation of the EEG as a monitor of the activity of large populations of neurons – well founded, although not claiming to be absolute truth, i. e. that the higher the voltage, the larger the number of neurons active at that moment; what we have called "synchronization". Since at the same time the higher voltage waves also exhibit lower frequency components, whereas the smaller ones exhibit higher frequencies, the latter situation has been called "desynchronization", meaning that a large number of neurons, instead of simultaneously developing voltage fluctuations, are acting asynchronously. The logical conclusion is that in the EEG it should be possible, in the case of extreme desynchronization, to detect small wavelets representing the "elementary" membrane fluctuations of individual cells. There is no trace in the reports of such a finding, probably for two reasons: on one hand, the smallness that can be attributed to such wavelets would require an amplification of the EEG also bringing to light the instrumental noise; on the other hand, the superficial and deep sinks and sources of currents due to simultaneous excitatory and inhibitory synaptic activations, would produce an annihilating effect on the elementary waveforms.

The problem of the quantification of continuous time functions fostered what has become a science within a science, EEG data analysis. The latter entails a preliminary choice of the philosophy of approach to electrophysiological data, depending greatly, in turn, upon some more or less conscious internal conviction of the investigator. The choice is between two different, though complementary approaches, the frequency domain and the time domain methods of analysis.

#### *4.5.2.1 Frequency Domain Methods*

The EEG exhibits regularities of frequency of the potential waves which have permitted us to identify characteristic "pictures", or patterns, by filtering the EEG in a number of frequency bands, with limits empirically defined, though corresponding to a handful of fairly differentiated states of brain activity (e. g. stages of sleep and of wakefulness). The regularity of appearance of the wave patterns, as well as their predictability and repeatability under given circumstances, has made the frequency of the EEG waves the parameter of choice for representing the known "states" of activity. This leads to nothing more than power spectrum analysis, in order to quantify the frequency components of the EEG. The spectrum being derived from the Fourier transformation of a covariance function, the treatment is statistical. Moreover, since the spectra are computed for detecting the changes of the activity in different situations, at different times or places, another statistical measure, the coherence function, is computed, a kind of correlation coefficient squared between the components of two spectra, a presumptive measure of the amount of similarity between the underlying processes.

The choice of frequency analysis is based on a tacit assumption of which, usually, one takes no heed, i. e. that the EEG waves have a "reality" of their own. This would seem to be substantiated by the repeatability and predictability of the frequencies in given situations. However, although these properties fostered the frequency analysis, they cannot be considered, at the same time, to be a test of the assumption. The latter seems to lose consistency if we think that the frequency of the EEG waves is the result of summation processes (synchronization and desynchronization) of excitatory and inhibitory influences, of sinks and sources of currents occurring simultaneously both on the surface and in the depth of the brain mass. For want of an unequivocal demonstration that EEG frequencies are not an outcome of the statistical interaction of all the membrane processes or, conversely, that they represent the product of a well defined operator on the activity of the neurons, frequency analysis cannot lead much farther than the comparison of different situations for their likeness. However, there is some evidence for the second hypothesis. I must recall the point stressed before, that recording the slower and graded membrane potential fluctuations is more strictly related to the input of the neurons than the recording of nerve impulses (which represent the output and are not in a linear relationship to the input). Therefore, the EEG from any area of the cortical surface does not represent simply a monitor of the activity of the local neurons, but of the activity at one point of a circuit which includes, as well as the cortical, those subcortical neurons (for instance, of the thalamic nuclei) which maintain the more strict and direct connections with that area. This fact is not at all new. It was recognized approximately 50 years ago, but seems to have been forgotten by most neurophysiologists. The importance of these circuits in determining the pattern of the EEG waves seems to be recognized only with reference to the so-called "spindle" waves of sleep, a rhythmical activity originating in the thalamic pacemaker neurons directly linked with the cortical cells, whereas there is absolutely no reason why these strict anatomical and functional binds should not be considered as the determinants of the ongoing EEG patterns in every situation or state of the brain. Only from this point of view can the frequency analysis performed on the surface brain waves be physiologically justified, being not simply an analysis of the cortical waves themselves, but of the physiological operator underlying them.

#### *4.5.2.2 Time Domain Methods*

With this approach the EEG waves are taken on their face value, i. e. they are simply considered as continuous voltage-time functions. What one usually does is to sample the continuous functions at fixed intervals, thereafter treating the samples as a stochastic time series, which means the application of statistical methods of data analysis.

There have been successful attempts at extracting information from the overall activity of a population in particular situations, e. g. when recording with thin wire electrodes in large populations of nerve fibers, where the individual nerve impulses are no more discriminable, and the recordings have all the characteristics of a continuous time function. In that case, the time function is not sampled, but can be averaged in different ways. Provided the nerve impulses within the bulk of fibers are randomly distributed in time, and their average does not change with time; that is, provided the system is in a steady state, an average measure, such as the root-mean-square voltage, is directly proportional to the average number of impulses in the territory of the electrodes. The method has been successfully applied to measures of the activity in the optic nerve and in central fiber bundles, the pyramidal tract, and optic radiations, but is not suited for recording from populations of neurons, on account of the simultaneous presence of differently oriented sources and sinks of currents. We shall see next (Sect. 4.5.2.4) that the information content of the continuous time series can be utilized through a procedure which is only slightly more elaborate, but is applicable to all types of recordings and physiological situations.

#### *4.5.2.3 Cortical "Evoked Response" Potentials*

Let's turn our attention to the most popular method for quantification of the activity in special situations, which has become one of the principal tools of investigation in neuroscience: the "evoked response" method. The surface recording (i. e. from the brain cortex) of the activity of large populations of neurons has had its best exploitation in evoked potential recordings. This practice has been considerably refined in the course of the last three decades, thanks to the refinement of the processes of data analysis. Potential oscillations on the cortical surface are evoked by sensory stimuli (in some cases also by electrical stimulation of nerves or of central structures). The principle is that by summating the potential waves (evoked by a number of regularly spaced stimuli) at fixed intervals after the stimulus, one obtains a voltage increase of all the potentials evoked by the stimulus, which are obviously falling in phase at those intervals.

The potentials, which by definition are considered as not belonging to the response, i. e. as "noise", being decidedly out of phase, exhibit different polarities from time to time, and thus are progressively reduced. There is an improvement of the signal-to-noise ratio proportional to the square root of the number of the summated potentials (of great advantage when recording from the scalp, since the EEG waves amplitudes, as we have seen, are in the $\mu$V range). What one obtains, however, is a complex of potential waves, or wave pattern, representing an "average" of the individual evoked responses, something from which has been eliminated the fluctuations of waveform present in each evoked response at successive times, therefore something representing a kind of "ideal" evoked response.

All this can be accepted only if one assumes that the variability of the waveforms in each complex evoked by a stimulus is exclusively due to real noise, a factor which is instantaneously unpredictable in as much as it is perfectly random, a rather unacceptable proposition. Noise, as a real, physical entity, is not, most probably, the principal cause of fluctuation of the waveforms, as indicated by the fact that the form of the single waves of the complex is strongly dependent upon their delay with respect to the stimulus, being more stable and repeatable the more the delay is short, and vice versa. The reason is that the potential waves at these short intervals are evoked by the arrival of impulse volleys from the receptors (or from other articially stimulated structures) through the more direct pathways, that is, through conducting systems in which there are to be expected the smallest dispersions in the time of arrival of the impulses. (There is more synchronization at the end of the stretch.) Conversely, the later potential waves exhibit a larger variability, because the impulses arriving later are those which have traversed a larger number of synapses (thus accumulating synaptic delays, and suffering a larger dispersion in time), and are more mixed with other nervous information not belonging to the response. Neurophysiologists have taken advantage of this, in order to discriminate: i) through which pathways the impulses are arriving, and ii) from the analysis of the waveform changes (shape and amplitude) occurring in physiological and in pathological situations, where in the neuronal circuitry the corresponding changes have occurred. It must be remembered that the very repetition of the stimuli brings about an improvement of the stability of waveforms due to the summation at the neural membranes of the effects of the successive stimuli at fixed intervals, thus reducing the amount of unpredictability in the synaptic processes following a single stimulation, as we have seen in Sect. 4.4, a fact independent of the summation of the potential waves in the evoked response method. The evoked response recording, processing and quantification allows only a statistical treatment, mainly aiming to analytically explore the nervous function in a circumscribed area. While in the raw EEG it has not been possible to detect the elementary wavelets representing the elementary synaptic processes, in the "ideal" and elaborated waveform of the evoked response complex it is possible to isolate "ideal" elementary waveforms of different shape and voltage, the linear combination of which accounts for the results of the analysis of the wave complex. Again, this is the product of applied statistics, in this case factor analysis, which has shown its usefulness beyond any doubt in neurological diagnosis, in assessing drug activity as well as in neurophysiological research.

#### *4.5.2.4 The Method of Correlation Between Brain Activities*

In all the methods in which statistics is involved, either frequency or time domain methods, one always has to take heed of the imperative constraint to which the methods are subjected, i. e. the necessity for stationary conditions (in the case of the brain, for steady state situations). Particular care must be taken in evoked response research, since the repetition of stimuli, (the largest feasible, in order to assure the largest gain in signal-to-noise ratio) may require several minutes, and during this period the conditions are like to change, not simply because of the natural evolution of the brain activity, but also because of the very presence of the stimulus. However, the same is true for whatever method of quantification when applied to long term analysis of the nervous activity. Therefore, we need a method in which the requirement for steady state conditions is relaxed, or altogether abolished.

Without the evoked potential analysis, processing the electrical activity of large populations of neurons with the usual time domain methods would seem a hopeless task. Only during non-attentive states and the in earlier stages of sleep do the time functions exhibit waveforms and patterns recognizable in different brain areas (although not everywhere and with variable phase lags from point to point). In the other situations, the fine patterning of the EEG does not offer recognizable waveforms, and the time functions are different, in general, even for electrodes a few mm apart from each other, that is, in immediately adjacent areas. Therefore, apart from the situations in which waveforms can be clearly identified, there would seem to be little or no hope of extracting information from the voltage of the EEG waves, but there are ways out. The "territory" from which the surface electrodes are picking up the activity is at least as large as the contact area of the electrodes (it is, actually, somewhat larger). The neurons of that area are not likely to be synchronized, otherwise well-defined waveforms would appear in the record, so, in order to be able to compare the activities of different brain areas on a quantitative basis, it becomes necessary to average the recorded voltage oscillations. Obviously, the averaged potential from a large electrode is the "time" average of a "phase" (or space) average, thus the time constant of averaging must be short enough to assure that the conditions of steady state are preserved, since the meaningfulness of the averaging would be completely lost, were the steady state conditions not respected. Time constants averaging 1 or 2 s are generally considered to be quite safe under this constraint, and the continuous "integration" performed electronically (or digitally) is accepted as quite a suitable average. Certainly, with this procedure the information related to the phase of the individual brain waves is lost, but it turns out that the time patterns of the integrated functions from nearby and faraway areas exhibit strong similarities, i.e. they are in phase during long periods of time, even in the absence of any recognizable waveform in the raw EEG record. Under these conditions, the other time domain methods fail, having to deal with completely desynchronized patterns.

The problem of giving a quantitative significance to the integrated voltage patterns of the EEG cannot be solved through absolute measures, but only through relative measures in arbitrary units, by comparison with similarly integrated voltages of unitary impulse discharges, or with integrated electromyographic (EMG) activitity. This is not, as it would seem at first sight, a second-choice solution for want of a more precise and better method of measurement. Actually, it corresponds to the fundamental requirement of brain investigation, the principle of comparison of the activities of different structures, and of structures of a different nature (such as the EMG, movement itself, changes in secretions, in electrical impedance, etc.). As I have maintained in Sect. 3.2, the principal factor on which we must base our knowledge of the brain is represented by the relationships between structures, because it is not so much the increase or the decrease of the activity of this or that structure that matters, but a definite pattern of activity of each structure in relation to the activity patterns of all the others. Organization, the fundamental property of nervous function, is expressed by the relationships between the different parts. Since in the study of the different structures of the brain one cannot be limited by the type of electrodes, their size, or their position, it is impossible to appreciate all the recorded time functions of the activity in the same units of measure, unless one resorts to an arbitrary, unified and relative measure for all. Following this line of thought, the problem of the exact measurement of the activity in the nervous structures loses its urgency, because the methods of comparison can easily do without it, in as much as they determine the extent to which the change of one of the measurements is related to the change of the other. Again, this may give rise, though not at all necessarily, to a statistical treatment, for which, however, non-statistical procedures can be better substituted, leading to much farther reaching conclusions than are usually provided by the statistical analysis of variance. Practically, a statistical comparison can be performed through the computation of the cross-correlation coefficients between pairs of variables, thus obtaining in the end a matrix-like array of the coefficients defining at each instant the state of the relationships between all possible structures of the brain. Recently, the method has been submitted to testing with a reduced number of variables. Later on, we shall meet the other, non-statistical procedures.

A final comment on the two types of investigation as they have been identified in the preceding pages. What we have called "analytical" research, mainly, if not exclusively, performed through microelectrode recordings of single neuron activity, aims to identify the structural and functional organization of circumscribed nervous circuits. In my opinion, however, the great many results gathered by tracing connections at the single neuron level, and by describing many of the functional properties of single cells in different structures and for different types of physiological stimuli, cannot lead directly to knowledge of the essentials of "organization". The latter being the principle upon which brain function is founded, knowledge of it cannot be obtained through the painstaking, piecemeal gathering of information from single cells. It would be possible to do so only if something of that principle of organization was present at the level of the single neuron, but then it could not be reconciled with the other principle, the very essence of the concept of organization, i.e. that the properties of the whole transcend the sum of the properties of the single parts. Analytical investigation can reach a most detailed description of connections and of individual properties, deep in the machinery of structure and function, but cannot by itself retrace its path upward, toward the general principles which direct the function of the whole. On the other hand, the method of investigation we may call "integrated", the study not only at the level of single neurons, but mainly of populations, and dealing principally with the relationships between parts, is not simply complementary to the former, but, striving to detect the relationships between all the elements, aims more directly at the core of the problem "organization". How this can be accomplished has not been made clear in the preceding pages. This is the problem to which the rest of the book is devoted, needing a long theoretical discussion of the physical properties of this organized object, our brain.

# 5. Multidimensionality, Homogeneity and Fields

## 5.1 Multidimensionality Versus Unidimensionality

We have seen in the preceding chapter how the investigation of the brain has enjoyed so far a number of technically differentiated approaches, and that, in the very end, the philosophy of research follows one of two lines, the analytical or the integrated. One aspect of brain function cannot fail to strongly condition the attitude of the investigator: the great number of neurons are organized so as also to maintain unity of action in those situations where only a limited number of the brain subsystems seem to be involved. Corresponding to this particular property is the fact that behavior results from the interaction of all the brain subsystems, and must be regarded as the unitary product of the unceasing process, or response, of adaptation to continuously changing conditions. If we want to investigate how this process develops within the brain, we are forced to study the simultaneous activity of all its parts and their mutual relationships. Thus, we have a problem of multidimensionality, offering several aspects in turn. On one hand, the ten thousand millions of neurons constituting the brain potentiality, each of which has as yet undefined degrees of freedom, either individually or in groups represent the "variables" of brain problems. On the other hand, the innumerable working conditions to which the variables are submitted from time to time contribute to increase the multidimensionality of the system. These two aspects coincide when the analysis is carried to the single neuron level. Almost certainly the fine details of the connections of any two neurons will exhibit differences, and thus every variable actually has different working conditions from the others. The difference between their individual working conditions determines the different behavior of the neurons.

The unity of action exhibited by the brain prevents a simplification of the problems by circumscribing attention to limited areas, or to apparently simpler functions, such as the reflexes. In any case, the number of variables would be reduced from the ten thousand million level to millions, or even thousands, however there would always be a large number. It is much better, then, if we want to continue with an integrated approach, to get used to thinking in multidimensional terms. After all, this is quite common in mathematics and physics, where these types of systems are dealt with through suitable mathematical procedures. This way of proceeding is not easily accepted, though, since the biologist is afraid to get acquainted with procedures and techniques too far from his immediate interest. However, one is always aware that living systems are multidimensional, a reality haunting the investigator who steadily seeks a way out.

One of the main hindrances in dealing with many dimensions is that the mathematics involved is complex and has been developing rather slowly to meet particular situations, thus restricting its usefulness to a limited number of cases. As a consequence, investigators revert to trying to simplify the task by dealing with one variable, or one condition, at a time. Experiments are set up either for the study of the effects of a number of different conditions on one single variable, or of a single condition on a number of variables. The extent to which this approach works is determined by the degree of independence of one variable from the others (or of one condition from the others), so it evidently cannot be equally suited to all variables, or to all conditions. Working with only one variable means letting it take on all possible values, which can happen only if the conditions of the experiment are allowed to change. To the extent to which that variable is independent of others, the observed changes can be attributed to the effects of the changing conditions on the variable itself. Otherwise, the changes are only secondary, and the result may be altogether meaningless or trivial. That this may occur rather frequently can be easily assumed, considering that if one variable can change, there is no reason why others (and in large numbers) should not change too. On the other hand, there is not much to be gained by trying to fix the experimental conditions: the possibility of controlling the conditions is limited to the external ones, and it is extremely difficult to keep them constant if they all must be really controlled. As for the internal conditions, these are altogether unknown, with the exception of a few vegetative parameters (such as blood pressure, oxygen and carbon dioxide content, etc.) and of a handful of variables correlated to the one being studied. Since this applies to all the internal variables which make up the ensemble of the internal conditions for the variable studied, the procedure is almost pitiful. Therefore, this approach, that we may for simplicity call that of the "one variable", has in practice produced many a failure because of the impossibility of maintaining the constancy of all the variables but one; both elements, variables and conditions, are so strictly related that a change in one brings about a change in the others. Nonetheless, we cannot say that experiments utilizing this approach have not produced important and reliable results, the reason being that one takes heed, in the course of the experiment, of changes in conditions and variables in order to explain the variability of the outcome. One cannot deny that this has been a reasonable procedure: from Claude Bernard's time up to now the textbooks have been based on data collected mainly, if not exclusively, using the "one variable at a time" philosophy.

## 5.2 Can We Work with Simpler Systems?

The drawbacks of the one-variable method are always lingering in the subconscious of the neurophysiologist, but he keeps thinking that somehow his problems can be simplified, either by effectively reducing the dimensionality of the system (which seems doubtful, to say the least) or by some additional assumption to assist the fundamental one-variable principle. The easiest way is to make additional assumptions, usually concerning the variability of the variables. Mostly, what the experimenter likes to think is that "on average" the general state is not changing with respect to all the variables but in the one examined, at least for the short interval during which he makes all possible observations (measurements) of that variable. This assumption, of course, does not avert the objections to the one-variable principle, but does to some extent relax the constraint of constancy of conditions in the general state of the system, simply requiring that the oscillations of the parameters of the external conditions, and of the values of the innumerable internal variables, do not alter the "quality" of that general state. Everything then rests on the definition of the quality of a state, which in turn depends upon a number of parameters (the variables themselves) used to identify it. (For the definition of brain states see Chap. 7.) The smaller the number of the parameters (variables), the lower the precision of state definition, so that, with the tendency of the investigator to monitor the smallest number of variables (which is at the basis of the multidimensionality reduction policy) one is never sure that the general state has not changed, unless in a roughly approximate way. It must be also remarked that even the use of a handful or more of variables to define a state does not necessarily mean that one is working with a multidimensional approach philosophy, since those variables are simply taken as an index of the situation, whereas a multidimensional approach is essentially a study of the "covariation" of many variables (but can also be performed on a restricted number of them).

The more we consider the real working conditions of our brain, the more the hope of effectively reducing the dimensions of the system becomes vanishingly small. The fact is that in everyday life our brain is continuously urged to respond in a correct way to a host of ambient stimuli, constituting with their many varying aspects, a "complexion" of conditions, that is, a different situation at every moment. There is no actual, reliable way of experimentally reproducing that ideal state of change in only one variable, simply because we can control neither the external, nor the internal conditions represented by the variables themselves. Moreover, the variables are time-dependent processes in evolution which cannot be interrupted, unless perhaps in artificial and extreme conditions of doubtful value and interest, such as, for instance, states of deep anesthesia or coma. In a way, the latter situations could be considered as means for somehow simplifying the system (reducing the number of variables), like other experimental procedures used occasionally (isolation of portions of brain tissue, or studies of neuronal properties in tissue culture, which have recently produced valuable results). To the same category belong attempts at finding analogues of brain circuitries in simplified models of a few neuronal connections in invertebrate species, or in nets of artificial neurons designed with specified and "ad hoc" properties.

The only way of reproducing a much less unnatural complexion of conditions, one which also easily occurs spontaneously in nature, is the suppression of one of the conditions, the one of which we are studying the effects, leaving all the other natural factors as undisturbed as possible. This approach has made its fortune particularly in neuropsychology, where the external conditions can be manipulated to some extent (examples dealing with maturation and learning have been reported in Sect. 3.3). Transferring the method inside the brain mass, with the aim of suppressing internal conditions, is not so straightforward, the internal conditions being mainly represented by the variables themselves, the suppression of any of them would mean, practically, damaging the brain structure, with consequences that cannot be referred simply to the absence or deficit of some factors, even less that of a unique factor or of one variable (in Sect. 4.1 some considerations on this point have been put forward). Moreover, the very appraisal of the amount of change in some function is limited by the availability of sufficiently sensitive tests, based necessarily on either a reduced number of parameters (measurements of a few internal variables) or on behavior. The latter, although representing the outcome of all the variables and conditions, does not allow direct, or easy measurement of them, and in itself represents a unique (though complex) parameter of a qualitative nature, which needs to be broken down into its multifarious elements in order to be quantitatively appreciated.

Certainly, working with more than one variable and one condition at a time requires a more elaborate experimental design, and refined methods of collecting and analysing data, but there is no doubt that it reduces the time otherwise spent painstakingly gathering results from the largest possible number of experiments with the one-variable method. Fewer experiments, in which the largest possible number of variables involved in a phenomenon are monitored under a wide gamut of varying conditions, produce much more valuable and reliable results. Being thoroughly convinced of this, I think that the one-variable tenet of experimental research in neuroscience should be abandoned in favour of a multivariable and multi-varied conditions philosophy. After all, our psychologist colleagues are already accustomed to working with as many variables at a time as feasible, having to deal with answers to stimuli of complex informational content, and having to account for individual, interindividual, and population variables, and all the other factors influencing brain behavior. We are in debt to the psychologists for having introduced the systematical use of factorial analysis, canonical analysis, multiple regression methods, and others, all suitable for dealing with many variables at a time, thus helping to make data analysis a really new science within the neurosciences. Data analysis may seem to have progressed slowly until recently, because of the computation burden inherent in the methods, but the present availability of automatic devices has enormously facilitated the task, and there should be no restraint in its extended use.

## 5.3 The Problem of Homogeneity

Many problems would be enormously facilitated, had we only to deal with homogeneous systems, i. e. if the components could all be considered to be fairly similar as far as their functional properties are concerned. After all, the basic physiological properties of neurons, individually considered, are very similar, differing, perhaps, as to the threshold of excitation (a parameter which would also affect their maximal rate of discharge). If the reduction of the brain's dimensionality could be achieved on these grounds, it would entail the consequence that, in a way, the neurons could be treated as numbers (rather like the molecules of a fluid). While this could be accepted to some extent for fiber bundles, as has actually been done (see Sect. 4.5.2.2), it represents an unacceptble proposition when neurons are involved, their activity being thoroughly conditioned by the anatomical connections within their circuit. This means that at every moment the activity of each neuron is differentiated in a way depending upon its position within the circuit. On the contrary, molecular-type homogeneity would imply that, as with the molecules of a gas, there could be an interchange, functional, if not surely anatomical, of the neurons, without a change in the properties of the ensemble. Of course, this is a proposition "ad absurdum", but it does not exclude beforehand that a reduction of the dimensionality might be reached by assuming that subsystems (centers, nuclei, groups or less anatomically circumscribed neuronal pools) could be considered, in some way, as being homogeneous in as much as they are composed of neurons having some (though, certainly, not all) common functional properties. The basic rationale for grouping the neurons in classes, one of the main products of analytical investigation, is that such common functional properties for building homogeneous classes exist. As we have just seen, a detailed exact homogeneity should be excluded at the level of single neurons, there being no exact one to one correspondence of the connections binding any two individuals, but it could be accepted to a reduced degree at a somewhat higher level of anatomical organization, that of a group considered as part of its ensemble. We might call it a "functional" (or physiological) homogeneity. It must be clear, however, that the homogeneity so defined cannot be a characteristic fixed one for all, since it might not persist unaltered when that small ensemble is involved in a somewat different function, and, very likely, would disappear altogether for functions performed at different levels of organization, since the latter would require the neurons to be differently sorted out.

This might seem to be a futile discussion, but it is not. A reduction of the dimensionality is badly needed, if not on theoretical, certainly on practical grounds, because of the intrinsic limitations of the instrumentation involved in the research, and of the real difficulty in handling the data, even with the help of powerful and sophisticated automatic systems. A reduction of dimensionality through homogeneous subsystems would bring with it the possibility of submitting the neurons to the statistical and thermodynamic treatment of the ensembles, the same, as we shall discuss at a greater length later on, which makes possible the investigation of systems with "ad libitum" number of particles, like fluids in general. Therefore, even if one cannot think of interchanging the brain neurons, the acceptance of some degree of functional homogeneity would permit us to substitute for a large number of neurons a few physical functions of them, in the same way that the thermodynamic variables are functions of the ensemble of molecules. (We shall talk about this in the next chapters.)

## 5.4 The Hypothesis of the Continuum

There is another way of looking at our problems. It is apparent that the concept of functional homogeneity must have limits and that these are generally set altogether arbitrarily. To what extent the neurons belonging to one of the classes can be considered homogeneous, depends upon the precise definition of the properties used for forming the class, and upon the permitted range of their variability, since, on account of the unavoidable differences in the individual interconnections, a variablity of behavior of the same, and of different, neurons in the same situation will result. It might seem, again, a matter of statistical definition, and it could be if we stick to the concept of class, but it is not necessarily so. The basic idea of a class is that its members, being functionally homogeneous, are more or less alike and their activity can be submitted to the usual statistics which treats them as numbers, and not as individuals each one with its own specific functional properties. If we want to preserve the latter feature (the individuality) of the elements of the system, and still be able to handle them in a unitary way, we must introduce other concepts. The more useful one, borrowed from physics, is the concept of "continuum", that of a system (material or mathematical) occupying a space, every region of which, however small, contains elements of the system. A continuum, then, is a system which is without discontinuities (look, however, for the concept of "singularity" when discontinuities appear in some parameter of the continuum) although not necessarily homogeneous in the physical sense (density of elements per unit volume) since the elements might not be evenly distributed. Also, the properties of the system need not be equal in all the regions of the space occupied, i. e. the continuum, in physical terms, might be anisotropic. What I am putting forward here is not an analogy. On the contrary, I am actually proposing that in our system (of neurons, of properties or of functions) the elements are building a nonhomogeneous continuum, and that the differentiation of the elements is adding anisotropy to it. This concept helps us to better define what had been anticipated in Sect. 2.6, i. e. that the activity of the neurons can be viewed as being originated from the interplay of forces, and that the forces build "fields". Now, the continuum is the medium where the forces build their fields. The continuum "is" the field (or many fields).

In order that this proposition should not remain mere speculation, we ought to find out how it may obtain the necessary confirmation. So let's see, first of all, where it is leading to, by supposing that the conditions of continuum are met within some brain area. If we are not willing to admit the presence of nonhomogeneity and of anisotropy, then the field of force will be perfectly homogeneous throughout. Since the forces represent the influences exerted by the activity of each neuron upon others, a perfectly homogeneous field can only mean that there is nothing to differentiate the neurons from each other, i. e. functionally they would all be identical, and the whole ensemble would be active as a homogeneous "mass". In other words, as far as the effect of the impulse discharges is concerned, all the neurons would be altogether undistinguishable, which is again, although somewhat relaxed, the proposition we were not willing to accept in advance (Sect. 5.3.), the one which would make of the neurons so many "molecules". But, since we can affirm that each neuron has an activity differentiated according its own position within some circuit, then the continuum is actually anisotropic, and in the fields of force are built, thus, patterns of lines of force, which differ from time to time following the changes of the activity of the elements: the field may be said to be "structured", or to build "structures" of forces.

However, it seems that the two properties, isotropism and anisotropism (unstructured and structured fields) may coexist in the same continuum. There is universally accepted experimental evidence (collected a few decades ago) related to the activity of the retinal ganglion cells in different conditions of retinal stimulation. In the presence of patterned visual stimulation, these ganglion cells behave in a well differentiated way, following the characteristics of the stimulus falling on their corresponding receptive area. On the other hand, in the limit condition of unpatterned stimulation, and much more so in darkness, the behavior of the different neurons is no longer differentiated. In the first instance, patterned visual stimulation, the central effect of the optic nerve discharges is to rouse a visual perception (of the patterned stimulus). In the second situation, the several millions of impulses per second coursing along the optic nerve, besides generating the "simple" sensation of higher or lower diffuse luminosity, have been shown to also contribute to the brain's "central excitatory state", thus exerting an undifferentiated "energizing" effect upon the CNS, what we shall call from now a "mass" effect. Thus, the retinal neuron continuum, depending upon the physiological situation, may exhibit either type of property, isotropism and anisotropism. If we take the example of the nerve cells of the retina as being applicable to other central neurons in the brain, we shall conclude that the presence or absence of structured fields within a neural continuum does not necessarily have to be a fixed characteristic, but is dependent upon the contingent working conditions.

Granted the presence of fields of force in the continuum of our central neurons, the basis of anisotropy and of field lines is that groups of neurons, or circuits of neurons in the continuum exhibit "preferential" open paths (what we usually call "facilitation", or "Bahnung") which endow the continuum with an organization, a structure in the functional sense, determining by itself the behavior of the whole pool. Thus, in each subsystem of our brain there may be field lines maintained by some mechanisms. About the latter, we may only advance hypotheses; a very popular one nowadays is that the different chemical properties of the synapses, forming systems of their own, and the "modulators", might be the factors maintaining the differentiation of properties within the continuum, its internal anisotropy. On the contrary, the neurons of an isotropic continuum, not being differentiated by a (temporary) differentiation of circuits (or of excitability) would, as a consequence, also not be differentiated by the characteristics of their impulse discharges. In the latter case, the field having no internal structure, the neuronal pools would behave as a "mass" in the physical sense. That is, the effects of their activity would be simply proportional to the sum of the effects of the individual components, as a gravitating mass may be said to exert a force proportional to the sum of its components, (though, as we shall see much later, the function cannot be linear on account of field interactions). As in the case of the retinal ganglion cells, the continuum of the brain cells may, or may not, be a structured field and have, or not have, as a consequence, a differentiated impulse discharge as a function of the working conditions, a property which can have far reaching consequences, as we shall see next.

## 5.5 Unstructured Activity and the Noise Problem

The fact that in a continuum with structured fields the activity of the neurons is differentiated, that is, the impulse discharges are structured, whereas in an isotropic continuum the impulse discharges are unstructured as well, tempts us to assume that the two types of functioning (mass and differentiated) correspond to the two types of activity the neurophysiologists like to call respectively "noise" and "signal". These terms are borrowed from that branch of physics dealing with the transmission of information, and have gained general use in neurophysiology in relation to investigations of the sense organs, chiefly the eye and ear. The steady, "spontaneous" impulse discharges of these sense organs and their extremely wide dynamic range of responsiveness impair the mechanism of transmission of information related to threshold stimuli. Therefore, the situation is similar to that arising in a communication channel where faint signals must be transmitted in the presence of strong noise.

The concepts of signal and noise from the sense organs have been extended to nervous activity in general, the concept of noise covering, little by little (and unconsciously), all "unwanted" activities, that is, those signals with which the investigator is not concerned at the moment. Of course, this becomes a blunder and a source of misconception if accepted outside the strictly colloquial sense, i. e. that the noise thus conceived is simply everything that masks the wanted signals. However, the concept of noise is much more precise and restrictive, because true noise represents time-independent processes, and intrinsically lacks internal correlation (as expressed, for instance, by the autocorrelation function), not simply a cross-correlation with some signals. On the other hand, the signals are generated by time-dependent processes and are characterized by an internal correlation of some sort; they may, or may not, be correlated with other signals, but cannot be cross-correlated with true noise. The measure of the correlation is a statistical probability; wherever there is a signal there is structure and information, i. e. decrease of disorder; the latter reaches its maximum in noise, and, generally, is measured as entropy in statistical terms. No reliable approach to the noise and signal problems in the nervous system can be based on a confusion of terms.

In the central nervous system too, true noise should be represented by an activity having the characteristics of physical noise, i. e. time-independent and uncorrelated with itself. That the nerve cells can originate impulse discharges with an apparently time-independent course could be demonstrated (only to a limited extent) by taking, for instance, time averages of the activity of single cells, but it is not easy to demonstrate that the apparently random impulse discharges do not conceal information, because the presence of faint signals, randomly occurring and with non-constant periodicities, would probably be missed. On the other hand, the assumed absence of information does not necessarily imply that the impulse discharge should acquire a random, disorderly character, i. e. that there should not be structure in it. Let's again consider the retina: its neural net, after inactivation of the associated receptors, can give rise to the most regular rate of impulse firing, then strongly autocorrelated. Only when the receptors and the neural net are intact, and not pattern-stimulated, does the production of nerve impulses seem to become random, like a noise. Conversely,the presence of noise-like activity of the neurons does not necessarily imply that the activity is not carrying information, thus representing a signal. The average level of the random retinal activity (see the preceding section) is inversely proportional to the average level of illumination of its visual field, provided the latter is unstructured and diffusely illuminated throughout. This shows that the concepts of noise and signal cannot be transferred directly to the nervous system, but also indicates that in some way the central neurons are able to "understand" not only the information carried by the structured activity, but also that carried by noise-like activity as well.

Since in each nerve fiber the structure of a signal is determined by the temporal patterning of the impulses, the correct interpretation of either noise or signal by the neurons should involve some measure in which time is the basic element. In other words, we are led to assume that somewhere there should be a "neural clock", and that the measure should in some way provide a correlation with the clock. However, talking of clocks, it is clear that we may have been led astray by the current knowledge about information theory and the artificial devices for detecting signals against background noise, whereas the nervous system may work on quite different principles. A hint is given, again, by the sense organs, their nerve nets, and their nervous connections. Whenever uncertainty could arise about the presence of a signal (faint signal and strong noise), what becomes evident is the phenomenon the physiologists call "lateral inhibition", the inhibition of the activity originated in the receptors (or in the neurons) surrounding the more stimulated ones, thus creating an artificial contrast. Although realized through a reduction of "unwanted" activity, the procedure would be similar in its effects to an increase of the signal-to-noise ratio (but, let me stress again, only if we also accept as noise those signals which in some way obscure the reception of the desired ones). Physically and mathematically, this is only a suppression of information, but it seems that, at least in the sensory domain, the nervous system adopts this procedure, which is much simpler to perform with neurons than a cross-correlation of some sort, not requiring moreover an undefined clock function. After all, having seen that the mass activity of a structure, which eventually may have the physical characteristics of noise, is actually interpreted by the receiving centers as being a carrier of information (about some steady state situation of a structure), we should conclude that in the nervous system there is no true noise, and that every neural discharge is carrying information, respectively of a purely quantitative type, the unstructured mass activity, or of a qualitative or mixed type, the time-structured and space-organized one.

### 5.5.1 The Processing of the Activity by the Brain

The double aspect of every nervous activity, mass and signal, does not decrease the dimensionality of the system. Since one cannot consider the physiological noise, or mass, as not carrying information, only its treatment could be, perhaps, performed with different methods. However, much more probably, a unique method of analysis will give the differentiated results corresponding to mass and structured activities. As I said previously, nowadays we are strongly biased in our thinking about the methods of analysis of nervous activity by artificial devices, not taking into account the actual quantitative aspects of neural problems.

Nervous structures are required to make decisions sometimes within a few tens or, at most, within a few hundreds of milliseconds, and during that time the number of impulses received by any nerve cell from any of its different input systems may be rather low, e. g. of the order of ten. While the impulses are being received, the cells should be able to discriminate, first, whether they are handling a mass or a signal impulse discharge (or both) and, in the latter case, the implied significance (quantitative, or qualitative). Therefore, the long averaging times necessary to detect the time-invariant structure of a noise are to be discarded, as well as the lengthy operations necessary to detect the degree of correlation and, thus, of structure of the message. The nervous system performs these functions differently or, rather, performs some other function with sufficiently analogous results. The cell membrane can integrate the discontinuous time functions (the impulses), generating continuous functions (the oscillating membrane potentials) which, in turn, generate discontinuous functions, and so on. After an impulse has been generated, there is a resetting of the integration period, the time constant of which, of the order of a few milliseconds, is believed to be a constant time characteristic of the membrane. For each sequence of impulses, and particularly in the presence of simultaneous inputs from many neurons, the arriving impulses can be integrated by the membrane (the sole process they undergo due to the actions of the membrane) only if they fall within the period of integration, and not during the period of impulse discharge. Operationally, this is similar to the "gating" function of a logical circuit. Such a gating can detect some of the characteristics of the time sequence of the incoming impulses, for instance, by revealing periodicities in the neural discharges, as well as temporal correlations between different inputs, without actually performing a correlation function. Thus, we do not ask ourselves any more by what methods the brain can differentiate mass activity from impulse trains carrying signals. It would seem that there is no need for two different methods of analyzing input impulses, since at the neuronal membrane the two types would represent both the resetting mechanism of the gate-reading function, on one hand, and what is going to be read, on the other.

### 5.5.2 Physiological Aspects of Mass Activity

The physiological significance of mass activity can probably be viewed as having at least two functional aspects, which are strictly related, although apparently different: i) the energizing effect, and ii) the gating and reading effect. The energizing effect was defined as the enhancement of the central excitatory state of the neurons, which results in the increase of their average level of activity. The phenomenon is demonstrated by the central effects of the retinal discharges in the absence of visual stimulation (the so-called "dark discharge"), and is thought to be mediated by those structures called "diffusely projecting systems", such as the reticular formation of the brain stem and the unspecific nuclei of the thalamus. (The diffuse projecting systems, though, effect a number of other more specific functions, since they are not anatomically or physiologically homogeneous structures, because older and newer parts have fused.) These structures, together with others connected to them, such as the hypothalamus, control the "state" of the nervous system by regulating the gross sleep waking cycle, and thus the general activity level of the whole brain, the level of consciousness, the flow of sensory information as well as the level of muscle tonus (the background level of contraction). Their mass activity seems to induce, or at least increase, the mass activity of other structures, a fact that may have profound functional consequences such as, for instance, the gating and reading effects mentioned before. The mechanism of synaptic gating, operated at some basic rate, allows only the information carried by impulse trains with the same or a lower basic rate to pass through the synapses; that is, allows the reading of information carried at some frequency of impulse discharge. This is the current interpretation of what is observed at several places in the nervous system, whenever two or more different inputs converge upon the same neuron. A typical example is found in the cerebellar cortex with the two input systems, the mossy and the climbing fibers. A gate controlling the sensory information is in the spinal cord at the level of the posterior horns, where the impulses carrying pain information can be blocked, or their transmission strongly reduced. Centrally, a gating-reading mechanism is certainly present in the unspecific and in the relay nuclei of the thalamus. The cells of these nuclei originate alternately (directly and indirectly through interneurons) excitatory and inhibitory post-synaptic potentials with the effect of opening and closing gates, thus alternately permitting and blocking the transmission of sensory informations. The mechanism, which is operated in large synchrony (thus at low rate) in the thalamic nuclei during sleep, is still present, though desynchronized (and thus at a much higher rate) during waking. As a general principle, every gating can be operated at different rates, but we can safely assume that there are structures so organized to reach an optimum level of performance only with a suitable level of background activity which, thus, must be maintained by the energizing effect of an "energizer". This might be, for instance, because the input signals are made up of sequences of impulses at high rate, which need a correspondingly high rate of gating in order to effect the reading of the signals with as little as possible loss of impulses. Cerebellum seems to work on high rates, while the thalamic pacemakers, as just said, are gating at low rate during the slow-wave, synchronized phase of sleep, and at a much higher rate during arousal.

I should like to make some short comments, to be resumed in later chapters, about the similarity of the effects of a neuronal mass activity and those of a gravitating, physical mass. The term "mass" for the nervous activity was purposefully selected, once its effects were better outlined, to convey the concept of that similarity. A gravitating mass is attracting other bodies; however, only by a stretch of imagination we could say that a mass activity in a cluster of neurons is "attracting" the same type of activity of other neurons. The similarity is not in this superficial aspect, rather, in the fact that both, neurons and physical bodies exert through their mass action an "organizing" function. Physical bodies through gravitation induce formation of orbitating systems and, on a smaller scale, gravity contributes to the differentiation of physical structures (but also the differentiation of the masses of the Universe is based for a large extent on the effects of gravitation). Mass activity of neurons, through the energizing effect and the mechanism of gating-reading information, can be assumed to be one of the basic mechanisms of the functional organization of the brain, certainly a source of differentiations, or of anisotropies in a continuum, since it can induce places of enhanced or decreased function in different brain structures, thus creating lines of force, i. e. field structures. Both, physical masses and neural mass activity are acting through the bulk of their constituent parts, but, far from giving rise to simply additive effects, have their action characterized by self-interacting phenomena (as we shall see later on).

Even if I have insisted on the continuum hypothesis as the playground of forces, it does not mean that our picture should remain necessarily a mechanistic one. However, in this book only the mechanistic approach shall be discussed. We shall see in the next chapter that there are other branches, as thermodynamics, in which the forces may assume a definite statistical character, and which would be equally well adapted to our descriptive purposes. The point I want to make is that the nervous activity can be described, in some of its most fundamental aspects, as the product of forces. A discussion whether they are better considered of a strictly deterministic, or of a broadly statistical nature, is left for later on.

# Part II Physics and the Brain

# 6. A First Approach: Statistics

## 6.1 Identification of a Framework of Laws

In Part I, I have briefly highlighted the main characteristics of the system "brain". The task we now are confronted with is that of outlining a theory of cerebral function in terms of the same principles governing the inanimate physical world. Attempts at identifying the physical principles which regulate nervous activity are not entirely new. In almost every neurophysiological paper mathematical or physical models have been adapted to the behavior of the variables under examination, and thus offer a description suited to the data. However, I would not like to follow this procedure, rather, I will try to broadly identify the domain of physics in which brain activity can be represented, and prove the identity of the laws pertinent to both the physical and the nervous systems. Thus, I will not look for models or analogues, but maintain the principle that the brain is only one of the many objects of Nature and is subjected to its laws. As was stated at the beginning, there is primarily a choice of the philosophy of approach.

Of course, in this procedure we should at first let the experimental data guide us towards a natural choice, but it is right at this point that what appears to be the most natural of the choices may not turn out to be the wisest one. Selecting a given physical framework implies that the brain is believed to be subjected to the general laws of that framework, and its functions to be describable with the pertinent terms. Physicists and mathematicians have developed theories, laws, and methods of investigation to describe the natural phenomena in the visible and non-visible world, but in our case the problem is not so much that of describing in an exact way how the brain behaves (after all, that would not lead much farther than identifying a model) but that of finding the explanations, the primary causes, and the ensuing chains of effects. The multidimensionality of the nervous system in terms of components, of functions, and of forces involved, and (primarily) the huge amount of data one collects in everyday practice whenever the brain is approached with measuring instruments, all lead quite naturally to the adoption of a statistical viewpoint. This, however, would not represent the choice of a philosophy, but simply of a tool, a method of describing and of analyzing data, as is always done with systems of many components of which one cannot follow the individual destinies. It seems the wisest, but it is only a contingent choice, which may or may not prove to be efficacious (therefore, wise and justified), depending only upon the nature of the phenomena at hand.

Actually, to start with a statistical approach would still require another preliminary choice, touching upon the basic problems of chance and determination, which come to our attention not as methods of description, but as principles from which the phenomena are derived. This was a problem bursting acutely only in subatomic physics. Since in the nervous system we are certainly not dealing with elementary particles, or with quarks, it would appear to be a rather farfetched reference to chance and determination in relation to the brain, were it not for what we have defined as its most important property, organization. Such a high degree of organization as that of our brain calls for more profound knowledge of the principles of its development than are available now, and for minds open to every possible solution of the problem.

### 6.1.1 Chance and Determination

The problem, though, is fascinating, and I cannot refrain from making a few observations. Whenever we have to do with phenomena whose occurrence seems to be dominated by chance, (such as the breakdown of atoms in a radioactive substance), the only reasonable way of dealing with them is to determine the probability of their occurrence, in order to be able to predict, to some extent, the future behavior of the system as a whole (unfortunately, not of the individuals). This is done through statistics, which has the sole aim of defining probabilities, and is the sole help to the scientist when he is unable to retrace the cause and effect relationships that are at the origin of each particular event. In this way, statistics has become almost synonymous of indetermination, or of chance, whereas being simply a descriptive tool cannot be charged with the burden of representing the mechanisms initiating a chain of events. But also beyond chance, the laws regulating the development of physical phenomena do not by themselves explicitly define cause and effect relationships, being generally represented by equations (such as, for instance, the equations of motion) which are perfectly reversible with respect to time, i. e. time-symmetrical. A cause-effect relationship assumes a precise time direction, from before to after, but physical laws may not contain prescriptions as to the direction of time, and hence cause-effect relations. What may give a direction to time are the "side products" of the phenomena described by the law, as energy losses from heat production, radiation, etc., which deprive the system of its capacity for reversing its "path" completely, that is, of obeying a symmetrical law. However, the laws of dissipation are "added" to the laws of motion, but do not alter their nature.

The fact that physical laws do not define cause-effect relationships, nor time directions, does not imply that they are not fully deterministic. Motion is fully determined by Newton's laws, the future position and movement of a body is predicted from kowledge of its present position and momentum. The validity of these laws is universally preserved, except at the level of atomic structure, where they break down. At that level, the physicists did not hesitate to abandon classical laws, and have created new mathematical tools (essentially, new statistics) and have been able to build a new body of knowledge, quantum mechanics. Fermi, Dirac, Einstein, and Bose have constructed a new statistics at the subatomic level, and we can only admire them. However, they are by no means the only examples. The beginning of this century has seen many other scientists adapt or create mathematical tools suited to their problems, such as the mathematics of field equations in general relativity, or the mathematics of matrix calculus in quantum mechanics. I think we neurophysiologists have much to learn from that approach. The epoch of very intensive experimentation which was initiated a few decades ago, during which our whole field was to be explored and all the methods were to be developed, should be considered closed. We should now adopt a more thoughtful approach to our problems. What is badly needed today is not so much a refinement of technologies, which seems unfortunately to be the general fashion throughout the world, as a refinement of the mental approach to research, particularly a widening of our individual cultural backgrounds. Of course, much time must be spent on the task, but this can only result in a better understanding of our own problems. Real progress is made only when we can put theories to the test, thus we should welcome any technical improvement, but efforts to produce it are justified only if they have been fostered by the real needs of the discipline.

Talking of the living being, the presence of cause and effect relationships is an assumption justified by what life itself implies, a chain of changes directed from the past to the future; a chain with no reversibility since at each step there are dissipative processes. (Life has been said to be founded on dissipation.) In the case of the brain, there is a progressive and irreversible decay of anatomical structure, sharply contrasting with the continuous evolution of intellectual power and the (apparent?) increase in stored information, which may go on up to the very end of the individual's life, provided there is a continuous supply of metabolic energy. Therefore, in the physical laws governing our brain, which should be essentially similar to those of non-living objects, we should also consider two requirements: the reversible formulation (as in Newton's laws) and the irreversible one (as in thermodynamics). For neural functions we ought to provide for some kind of reversibility, of symmetry, almost surely not to be accounted for in the same simple way as the reversibility of physical motion, therefore not contained in the same simple formulation. After all, as I have just stated, the true essence of brain function is not based on energy supply and consumption as it is for a non-living machine. It is true (see the next, and the following chapters) that there may be problems of energy measurement, but the question is specifically about the physical significance attached to the parameter "energy" in the central nervous system. Obviously, all types of nervous processes require the brain to be alive, but the brain's functions are independent, within the conditions limiting cell survival, of the energy supplied by metabolism. Thus, the dissipative processes, which in every physical machine prevent the conservation of energy and the exact reversal of motion, cannot, in principle, be of the same nature in the brain, and do not necessarily modify the symmetry or the reversibility of its functions.

On the other hand, while in the brain there is a temporally ordered succession of events, it may be utterly impossible to define in them a directed sequence of cause and effect relationships (apart from the trivial instance of well-isolated and temporally definable responses to stimuli), since they occur in an extremely complex maze of feedback circuits, upon which the structural organization of the brain is founded. The latter feature, together with the freedom from physical (or material) energy-dissipating processes, seems always to endow the brain's action with the general symmetry and reversibility we have been talking about. As I said before, the point is that we must not expect to see nervous processes reverse their course (like a motion in space) e. g. a nervous process leading to the analysis of the properties of a perceptible object cannot be reversed through the same neurons. However, the "judgement" given upon the nature of the object, the "synthesis" operated more or less simultaneously with the analytical processes is a kind of a "reversed" function, and may be fed back to the analyzing structures. However, that reversed function cannot physically be the mirror image of its origin, as reversed physical motions are, although both may be represented in a form denoting their reversibility, (e. g. a "transformation"). As a consequence, the laws of our brain, in their ultimate essence, ought to be represented in such a way as to represent definite types of transformations. I am perfectly aware that this interpretation of symmetry and reversibility is somewhat different from that contained in physical laws: no temporal reversibility is provided for, nor does the reverse action actually "undo" a previous action, (in other words, it does not nullify its effects). As for the first objection, the brain is in exactly the same situation as every other physical system; there is no actual escape from the directionality of time. The temporal reversibility included in physical laws stresses only their universal value as laws, so that a hypothetical reversal of the temporal directionality would still find these laws universally valid. This could also be attained for the laws of the brain, provided that in their formulation there is no recourse to time, or that more general formulations are used, in which time is not explicitly defining the before-after sequence. On the other hand, the undoing of what was achieved by a previous action is not feasible within the anatomical structure of the brain, on account of the impossibility of reversing the direction of synaptic transmission. Reversibility here can only mean that the "significance" of the function is reverse of another coupled to it. The interpretation I have just given of the symmetrical, reversible character of brain functions (as in the instance of analysis and synthesis) might not be found acceptable, and every brain operation might then be regarded as the result of dissipative processes, inasmuch as it is obtained through irreversible energy losses (whatever the nature of the energy involved).

## 6.2 Statistics and Neuroscience

As I was saying previously, organization, the distinctive feature of our brain, stands in the center of our problems asking our opinion as to whether chance or determination generated it. Although fundamental, this is one of those questions which are never settled, so in the following we shall go by roundabout ways with the purpose of avoiding a direct answer, on the one hand, and of exploring the wide range of possible solutions, on the other.

Let's start by considering the problem in a statistical context. Although, outside the subatomic level, the description of the physical world with the classical deterministic laws is sufficiently precise, nonetheless statistics has always had a substantial role in it. Statistics deals with populations of physical elements, or of data obtained from measurements, as is generally the case in neuroscience. Its basic principle is "homogeneity". This term in statistics must be intended in two ways: first, as we have seen in the preceding chapter, homogeneity of the properties of the components of the class, or population, from which the data have been collected; second, it is also clearly understandable as meaning homogeneity of a situation, a constancy of the conditions which the population is under, a time-invariance of some sort. The latter point is not always clear to the investigator who pays, generally, more attention to the characteristics of the population, but little, if any, to the constancy of conditions. It is true that the latter is not easily achieved in practice (and is difficult to monitor) nor, as we shall see, can it be clearly defined in every case. Statistics endeavours a calculation of probabilities, in the case of a sample that of belonging to this or that population, in the case of an ensemble that of being in this or that condition, of changing from this to that "state", and of developing in one or the other direction. The constraint of homogeneity cannot be escaped, lest the calculations of probability lose their significance. It all depends upon the fact that, for statistics, working with a number of identical "objects", the only meaningful measurements can be those based on averages, either simultaneous across a population (phase averages), or in timely succession (time averages). In the nervous system we are limited in the choice of the type of average by the fact that a measurement of phase averages requires us to simultaneously measure a great number of homogeneous variables: a procedure, in general, highly impractical or altogether impossible, so that, as a rule, we can avail ourselves only of time averages of one or a handful of variables.

In order to be able to draw meaningful conclusions under these circumstances, there is another principle which must be carefully preserved, the "ergodic" principle. What the latter implies is that, when sufficiently extended, the time average becomes identical to the phase average. This is the basic principle of statistical mechanics (see the next section) since it states that over a period of time any system in its "natural" motion will assume nearly all the possible values possessed by a large ensemble of identical systems measured simultaneously. Obviously, a formal proof of the ergodic principle is difficult to obtain in practice, and it has never been attempted as far as the activity of the brain cells is concerned, but the principle should nevertheless be assumed to be valid in the brain if we want to continue to base our knowledge only on time averages. The vast majority of neurophysiologists rarely take this condition into account and even seem to ignore it, while only its acceptance justifies the conclusions drawn from single cell experiments. In single cell recordings, it is tacitly assumed that the selected cell is representative of many other cells of the same class. (Actually, we take for granted beforehand that there is a class of cells with those properties.) In other words, the cell may be an "average" representative of all the cells of the class (a phase average). By recording from single cells, we can only follow their activity through time, considering the sequence of their impulses as a stochastic time series (produced by stochastic processes), which has meaning only as a time average, thereby introducing the problem of ergodicity. (Some physicists maintain that the very existence of an ergodic problem should be discussed, since practical demonstration of the ergodic principle is not feasible, but this is far beyond the scope of the present account and of the capacity of the writer, so I will just leave the question open.)

Yet, even granting ergodicity, the statistics of neurophysiological data require the system to be in a constant (stationary) or in a steady state, i. e. in a condition in which the state parameters are not changing with time. This cannot be easily determined, since long periods of neural activity are generally required in order to be able to ascertain whether the measures taken at different times actually belong to a time-independent process. It turns out to be a trial and error procedure, since one has to define: i) the length of the period of analysis, and ii) the permitted range of fluctuation (variance) of the measures within a given stretch of time. Thus, it all depends greatly upon the contingent situation, and the result is rarely foolproof. We see now that the boring discussion about homogeneity and the remarks about ergodicity were not futile, since they are aimed right at the core of the problem of statistical treatment. Only through well defined constraints does statistical data analysis become the more powerful method for symplifying a task which would otherwise appear hopeless.

The neurophysiologist, though making extensive use of statistics, rarely exploits its full possibilities (except, of course, for a few data analysts, whose achievements in the last decade have been outstanding). What one generally asks of the recorded time series is the type of stochastic process which has generated it (normal, Gaussian, Poisson, etc.), but all this offers little opportunity for meaningful conclusions about the physiological mechanisms, or the nature of the generator. As we have seen, the discharge of impulses by neurons is always the result of a complex interplay at the neural membrane of the inhibitory and excitatory influences of different origin. Thus, the impulse-generating mechanism is not operated upon by a unique process, but by an ensemble of different processes, rarely of a homogeneous nature, which cannot therefore be unified or considered in terms of averages. This essentially restricts the type of answers one can obtain from the statistics of neural activity to the statistical significance of the difference between the behavior of the variables (of their averages), the probability of occurrence of a given behavior, the possibility of forming clusters of characteristics, etc. In this way, statistics does not penetrate into the mechanisms of the physiological process itself, though being very helpful by saying "yes" or "no" (with defined probabilities of error) to a specific investigative proposition. The use of statistics as a tool actually means that one has resorted to an "a posteriori" methodology, useful for finding order in an otherwise unapproachable ensemble of data, but without any bearing upon the philosophy of the problems.

### 6.2.1 Statistical Mechanics and the Brain

Now we shall instead consider statistics, not as a tool, but as a principle which is at the basis of some of the important branches of physics. In this section our purpose is to find out whether statistics plays the same role in the nervous system as it has in physics. Statistics is the basis of four branches: statistical mechanics, first of all, then thermodynamics, quantum mechanics and information theory.

Statistical mechanics is the most general of the four, since it endevours to derive general mechanical laws which also (or principally) apply to systems with a great many degrees of freedom. These laws cannot define the behavior of individual elements of the system, but only of the whole. Newton's mechanical laws are strictly applicable to macroscopic rigid bodies; with fluids and other deformable media the situation is enormously more complex, so the need for a theory allowing the prediction of the behavior of any system was felt early. Statistical mechanics uses the same Newtonian movement parameters (position, velocity, and momentum) for each element of the system, and a convention, Lagrange's generalized coordinates, to build systems of differential equations which describe the motion. Although the convention and the terminology are specific for actual movements, i. e. for systems with moving parts, their usefulness is more, perhaps, for those systems with many degrees of freedom, not necessarily in actual motion, but nonetheless subject to changes with time (thus considering the change as a movement). As motion through physical space is described by actual coordinates of position, velocity, and momentum of a body, the system of equations based on the generalized coordinates of position and velocity defines the behavior of the ensemble of the elements of a system at any instant. For a non-moving system like our brain, the motion is better conceived of as a succession of "states" (the same, of course, can also be said of actual movement) defined by generalized position coordinates, which must therefore be selected so as to suitably represent the states (a problem in itself, as we shall see in the next chapter), and by generalized velocities (which are simply the first time derivatives of the positions). The system, then, is considered to be "moving" in a space with twice as many dimensions as the constituent elements.

Statistical mechanics essentially makes one type of prediction, those related to the "natural" movement (i. e., the time development) of a system when the latter is not subject to forces, which is to say the motion of a "closed" system. The "path" of a system in its natural movement has properties which principally involve the law of energy conservation, which projects the path to be covered, step by step, across surfaces of constant energy (as if the system were moving on the surface of a hypersphere). After moving about in this way, for a sufficient time, the path would cover practically the whole surface of the hypersphere, as would the instantaneous positions of very many such systems, simultaneously observed in their own natural movements. Thus, we re-encounter here the ergodic principle, briefly discussed above, which assumes that time and phase averages can be interchanged. However, here in statistical mechanics, being concerned with the path across a surface of constant energy, the ergodic principle also states that the overall energy of the system (its principal phase function) can be calculated through the time average of any summable function of the system, under some constraints, provided that the period of averaging is long enough to let the averages reach a limit (i. e. in strictly constant conditions).

With reference to the nervous system, the principle requires the components to form a homogeneous class in the strict sense, that is, of the "molecular" type. Therefore, only homogeneous groups of neurons could be involved, the existence of which has been admitted only in terms of classes of definable, but not fixed, functional properties, (not exactly homogeneity in the strict sense). Furthermore, the principle states that we can follow the behavior of these classes (through their time averages), provided their energy remains constant, a further constraint with respect to the first statement of ergodicity given in Sect. 6.2. The formal problem will be better understood later (see Sect. 8.2.2) when we encounter the Hamiltonian (a function of both position and momentum coordinates, equal to the total energy of the system). In the terms in which the Hamiltonian is defined, it entails constancy of the generalized velocities, but, for the brain, constancy of the velocities involved in the natural development of the state means that the system must be stationary. We should ask ourselves whether we can really expect the activity of the brain's components to be strictly stationary. In Sect. 5.3 the problem was first approached with regard to homogeneity of functions, which was not to be found throughout the stucture. Stationary condition was excluded in Sect. 5.4 by the first sketchy view of a continuum in which forces build their ever-changing fields, clearly in contrast to the specific requirement for a force-free situation (the closed system of statistical mechanics). A stationary state of activity could also be excluded by the discussion in Sect. 5.5, in which it was concluded that the activity of nerve cells always carries information, in a state far away from the stationary one, in which only statistical fluctuations of the activity parameters are allowed. All the same, I feel it is necessary to examine the matter more thoroughly, in order to clarify whether the brain could be truly stationary, if only under limiting conditions, or whether we are always confronted with steady states. The latter, are not exactly stationary in the way required by statistical mechanics for a closed system, i. e., a state not subject to forces. There may be forces acting upon the system; it is the balance of their effects which is null, or, alternatively, the differentials of the velocities.

A neurophysiologist is immediately tempted to identify a stationary state with "spontaneous" activity, and non-stationary one with "induced" activity. In the first instance, by definition, there should be no forces causing perturbation of the motion (the system is then left to its statistical fluctuations), whereas in the second, again by definition, the activity changes from its previous state to another by the intervention of some force (typically a stimulus), and the resulting motion is therefore time-dependent.

### 6.2.2 Spontaneous or Ongoing Activity

Many of the difficulties in identifying the physiological counterpart of the physical parameters of motion reside in the poor definition of the everyday concepts of neurophysiology. Spontaneous activity, by definition, should not be caused by any external influence on the neurons responsible for it. However, the term is used excessively to actually indicate activity in the absence of "intentional" stimulation, which is the only controllable type. It is certain, and tacitly admitted, that what we call spontaneous (we will call it, more aptly, ongoing) activity contains a proportion of neuronal responses to unknown and unwanted stimuli, unavoidable even in the most isolated conditions, since they mainly originate from within the body and the brain itself. The presence of such inexhaustible sources of stimuli raises the problem of whether among the different types of brain activity, anything actually exists that we may strictly call spontaneous. The only answer may come from neurons isolated by severing their connections with the rest of the nervous tissue. These experiments have demonstrated the actual impossibility of a neuron being active if the connections with at least part of its companion population are not preserved, as with isolated slabs of the cortex. The experiments, however, are not decisive, because the manipulations necessary for isolating smaller and smaller slabs may eventually result in damaging many of the neuron's branches, the satellite cells (glia) and the adjacent blood vessels supplying oxygen. Also inconclusive is the evidence obtained using cultures of nervous tissue, because of the uncontrolled velocities of the fluxes across membranes in artificial media. In retinas in which the receptor layers have been destroyed and, thus, the nervous net cannot be physiologically stimulated any more, the ganglion cells show what one would interpret as spontaneous activity (of a very regular type). However, the ganglion cells cannot be considered to be really isolated, being cross-linked through the amacrine cells in a still functionally unclear way, and may be subject to the even more unclear influence of centrifugal pathways from the brain. Thus, we are inclined to conclude that a minimum amount of intact circuitry is necessary to maintain a minimum level of impulse firing. In other words, a nerve impulse can be generated only in the presence of suitable input impulses to the neuron, and therefore consistently represents a response to a synaptic stimulus. The latter conclusion might not always be acceptable in such unequivocal terms. There are neurons in the thalamus, which have been named "pacemakers", after the heart self-excitatory pacemaker cells (those able to spontaneously and autonomously determine the heart rhythm). It is probable that the thalamic pacemakers operate as such by virtue of their cross-connections, and not by self-stimulation as the heart cells do. However, the question remains open. We have to accept as likely the possibilities that the individual neurons do not possess freedom, i. e. they cannot be active independently of their synaptic input, and that the only way of differentiating the types of activity in a neuron has to be based on the characteristics of the impulse discharge (e. g. time-dependent or steady-state).

In a steady state the generalized velocities are constant, but this can have meaning only in terms of a statistical average of the instantaneous velocities of one or of a population of cells. The averages must be allowed to converge to a limit, and this certainly requires long recording periods, subject to uncertainty about possible changes of state that may intervene during this period.

A steady state also implies that the level of activity is kept steady by constant operating influences. This is, after all, a normal occurence in the brain (recall the energizing effects of the high level of activity in the sense organs, and the diffuse projecting systems, discussed in Chap. 5 and also the next chapter). The conclusion again is that the motion obtained is not the exact counterpart of stationary, uniform motion of a body, because the latter implies that there is conservation of energy (conservation of velocity and momentum) and not dependence upon a continuous supply of energy. On the contrary, the ongoing activity in the brain is always the result of a sequence of short acceleration periods, which are generally undetected, if only because of the averaging effect of the large surface electrodes (as in the case of EEG recordings) or because they are obscured by the averaging measures necessary to detect the existence of a steady state (in the case of single cell recordings). A way out, or better, a compromise in order to allow us to distinguish something similar to a stationary motion from a time-dependent one, introduces an arbitrary standpoint into the decision. We could assume the motion in which one cannot detect these unknown, short acceleration periods building up the ongoing activity to be "quasi-uniform", on account of the large surface of the electrodes or of the averaging statistics used. An accelerated motion, on the other hand, would be assumed every time the acceleration is definite and directly demonstrable, as is typically the case with responses evoked by known stimuli with known origins in time. The introduction of arbitrary factors is a dubious procedure, but the decision is somewhat justified by the fact that averaging by electrodes or statistics obliterates the signs of short accelerations, whereas the same electrodes and statistics cannot cancel out the accelerations represented by evoked responses.

We could try to approach the problem of ongoing activity and stationary motion from a slightly different point of view. Since stationary motion is characterized by conservation of momentum, we could consider that type of activity which conserves energy as "physiological" uniform motion. At first sight, this standpoint seems to get us nowhere for two reasons: i) so far we have not mentioned any reasonable way of measuring energy. ii) we have just stated that even what we call "quasi-uniform" motion, is actually the result of continuous series of short accelerations, i. e. motion with a continuous supply of energy. Let's consider for the time being only the second reason. What provides the brief energy surges to the many brain subsystems, in order to maintain them in sustained activity, is simply another subsystem, which we know to be strictly connected anatomically with these subsystems through two-way paths, thus forming a unique circuit with them. One cannot distinguish a beginning, an end, a cause, or an effect, unless a "disturbance" is "injected" from outside. In other words, the accelerating subsystems must actually be considered as part and parcel of a larger subsystem which includes the subsystems accelerated. In this larger subsystem "neural" energy could be conserved, for instance, by being used as nerve impulses to accelerate a number of subsystems, and being returned from them, again as nerve impulses, thus, presumably, maintaining a steady state level of activity in the whole. These sustained (or tonic, in physiological terminology) influences have been always attributed to the diffuse projecting systems of the brain stem and thalamus, as abundant physiological evidence would indicate. This interpretation would be perfectly in accordance with the assumption made in the preceding chapters that subsystems are organized in a whole, in which cause and effect relationships are not clearly definable.

Accepting the relaxed condition, the steady rather than the force-free state does not prevent us from describing the natural motion in the nervous system with statistical mechanics. I have stressed that the brain's functions are independent of the (chemical) energy supply, provided the latter is sufficient to keep it ready to function. Thus, the brain behaves as a particular type of closed system, the energy losses from which do not alter its function. As I have just said, the brain's steady state, even if not representing true stationary motion, is still a conservative (energy-conserving) process, inasmuch as the forces maintaining the state's time-independence (brief, non-cumulative acceleration of activity) have their origin within the system itself, and are part and parcel of it. In this aspect the system should be considered closed.

### 6.2.3 Gibbs Ensembles and Steady State Motion

As the preceding discussion shows, a path covering a surface of constant energy would serve to automatically define a physiological situation (as long as the latter remains steady) and measurement of the energy under these conditions would amount to measurement of the energy of a state. Thus, statistical mechanics allows us to distinguish states of activity ("physiological" states) within the nervous system as well, provided they differ in total energy content. It follows that what we have called the states of the system in its natural motion, that is, the successive steps in the path followed by the whole system under stationary conditions, represent "elementary" states, practically indistinguishable from each others as to their energy content (being distributed on the same surface) and definable only through differences in the microdistribution of activity among all the neurons comprising the system.

It is not exactly prescribed through which functions the energy is to be measured; though, with reference to the Hamiltonian, position and momentum coordinates are in the first place as being representative of a state. The picture of the physiological, and that of the myriad of elementary states, will come out repeatedly and will be gradually better focused, as it represents something fundamental in the overall theory. Before abandoning the subject, however, we have to consider another point. Instead of an indivisible unity, we could assume our brain to be made up of an ensemble of subsystems with fundamentally similar general properties. A situation of this type has been considered by Gibbs, in the context of statistical mechanics, and has given rise, together with ergodicity, to one of the principles of that theory, the so-called principle of Gibbs ensembles (proved as a theorem by Liouville). Considering again the natural movement, now involving ensembles of similar systems, the theorem states that a constancy (a statistical one, of course) of the density of these systems is maintained throughout: the "constancy of density in phase" principle. This is assumed to hold for all natural movements, and associating with it the conservation of energy, the ensembles must move as an incompressible fluid. Transferred to the brain, the Gibbs principle loses much of its statistical connotation. The reason is twofold: i) The Gibbs ensembles are made up of systems which are all alike, i. e. cannot be distinguished from one another and are, therefore, "interchangeable". ii) The components of the Gibbs ensembles are not interacting. At variance with this, in the brain all the subsystems participating in Gibbs ensembles are certaily not precisely like each other (as is the essential assumption of statistical mechanics). On the contrary, it is most likely that they are all different (albeit in a steady state, in which each may preserve its energy constant). Moreover, the brain subsystems all interact strongly with each other. These extended interactions equalize the general level of activity among the subsystems, bringing the extreme velocities closer together, that is, inducing those similarities which define Gibbs ensembles. Furthermore, since in a steady state the differentials of the velocities should be null, the differences between the subsystems cannot preclude them from maintaining a constancy of relationships. Gibbs's reasoning, when extended to the properties of the brain subsystems, foresees for an ensemble of such subsystems, each in natural (steady state) movement, not only a constancy of energy and of density in phase (in this case, of a non-statistical nature) but also some constancy of relationships among the components of the ensembles, a kind of deformable-body movement. Why I insist on this point will be clearer if we consider that the brain must maintain a constancy of its basic functions (which are performed through precise functional relationships between the brain subsystems) under widely different conditions, i. e. in the same as well as in different physiological states. This subject will be resumed in the following chapters, where we shall again deal with constancy of functions from different standpoints.

## 6.3 Thermodynamics and Brain Function: The Onsager Representation

Of course, the statistical mechanics representation discussed in the preceding section cannot provide an answer to all the problems of brain function. A rigid constancy of relationships between the subsystems of the ensemble would become an unbearable constraint, since it would exclude any change when adaptation to different contingencies is required, i. e. in case of a "non-natural" movement, as is the norm for the brain's function. This is the point at which statistical mechanics, and with it the theory of Gibbs ensembles, reach a limit, because they cannot provide for the transition from one steady state to another, being concerned only with closed systems in equilibrium conditions. Perhaps extending the statistical approach to include the physiological states of the brain as defined only through their energy content requires a stretch of the imagination. Surely, by considering the natural motion with the simultaneous maintenance of constant relationships between the various subsystems, we have restricted the statistical connotation of the approach to stationarity of the steady state type. This aspect will be discussed again in later chapters, since the issue of constancy of the relationships between the brain's parts is at the core of every theory of brain function. Statistical mechanics fails to provide a suitable context for describing the transition between two physiological states, essentially because it does not make direct reference to forces which could cause the system to deviate from its natural course, whereas the action of a force, and some kind of exchange of energy are required when the system has to jump from one surface of constant energy to another, i. e. when work is done. Thermodynamics, though based on the general principles of statistical mechanics, also considers the forces involved in changes of state, since it can deal with "open" systems. Here, statistics does not necessarily, or explicitly, enter into the formulation of the relationships among the components of the system, rather, it remains hidden in the nature of the forces and, generally, of the state variables.

Dealing with systems as a whole, rather than with individual constituents, thermodynamics derives general laws of behavior for large systems with many degrees of freedom, typical of which are fluids. Thermodynamics does not consider a system in isolation, i. e., not as an ideal limiting case, but together with its surroundings, with which the system maintains exchanges (of matter, energy, heath, etc.). As a statistical science, its laws are mainly concerned, again, with states of equilibrium or with steady states. In living systems, a true equilibrium could be attained only in the absence of any exchange, a condition, as everyone knows, ultimately reached upon death. In the living, the only possible equilibrium is a steady state, whenever the derivatives of the changes, as revealed by measurement of the state variables, are null. During life there is a continuous exchange of matter (and, of course, related exchanges of heat, of electrical charges, etc.) between the cells and the surroundings. But, in a steady state there is also compensation for the inward and outward fluxes, balance of energy, and work done (including changes of information content). Through entropy, we balance gain and loss, taking into account the sources of irreversibility related to energy loss: in overcoming resistance, in friction, heat, radiation, etc. However, the two measurements, of energy and of entropy, are still of uncertain significance as far as the nervous system is concerned. The difficulty is mainly in determining which are the variables to be measured. The integrated activity of nerve cells and fibers, on which every measurement in the brain should most naturally rest, has never been accounted for in terms of fluxes of matter, or of energy in any form. Neurons generate impulses which, it is true, are representable as instantaneous changes of the ion equilibrium across membranes, and thus are based on fluxes, but which are never considered as amounts of fluxes, only as sequences of "quantized" events. Energy is physically bound to matter and movement, but in the nervous system we have so far made only a theoretical approach to the surfaces of constant energy covered by a system, and the energy has only been referred to as the principal phase function defining the system's states. Moreover, the latter have still been treated as abstract concepts, not yet linked to precise situations having a physical, or a physiological significance. The following chapters will investigate to what extent thermodynamics and other branches of physics may really help in this matter. The aim of the present and of the next section is more limited. We shall content ourselves with the general identification of the thermodynamical forces and fluxes within the nervous system.

It would seem that, after all, the thermodynamics of fluxes would not be very useful to the neurophysiologist, and that he should look elsewhere for directions enabling him to perform measurements of the parameters involved in the steady, and in the transitory, states of nervous activity. However, having developed from the basic principles of statistical mechanics, thermodynamics can actually deal with the most general types of flux. The simplest way of representing the factors involved is through the so-called Onsagers phenomenological relations, or equations, which give a general representation of the relations between the forces and the fluxes, both unspecified. A flux "J" equals the conjugated force "X" driving the flux, multiplied by a coefficient "L" which has the general significance of a conductance, (a typical example, of immediate physical significance, of this equation is given by Ohm's law). These phenomenological relationships are utilized to build systems of relationships between fluxes: a flux of matter is always coupled to other fluxes of, for instance, electrical charges, energy, heat, other material species, etc. This implies a number of flux equations, however not mutually independent, but coupled, the extent of coupling being defined by suitably chosen "L" coefficients. The latter, then, determine the extent to which any flux is driven by its conjugated force (the "straight" coefficient) as well as by other forces too, which are driving other fluxes (the "coupling" or "mixed" coefficients). Thus, whenever a flux is coupled to others, the resulting system of equations exactly defines the physical situation, however complex it may appear. The extreme usefulness of the Onsager equations as representatives of the interactions between the different factors and elements of a physical system is immediately apparent. Of course, giving a value to the coupling coefficients is the critical point in this representation, but there are rules by which the problem may be simplified. Although primarily conceived for physicochemical systems, the general form of the phenomenological equations can be applied to very different systems, including the brain (for which still we have to determine the counterpart of forces and fluxes). The reason is that the Onsager equations formalize an obvious, most general principle, that to obtain an effect (a flux, in the specific case) one has to use energy (apply a force) and that the result depends, in quantitative terms, upon certain conditions represented by the coefficients. We might say that the Onsager representation has the same role in thermodynamics as the generalized coordinates in Lagranges's and Hamilton's equations of movement.

Leaving aside for the moment the identification of fluxes and forces, let's consider only the principal characteristcs of the brain, the capacity for reacting as a whole in even the most simple situations. Each of these brain "reactions" comprises different "partial" functions (sensory, rational, emotive, motor, secretory, etc.) which in turn can be conceived of as comprising different subfunctions, all coupled to each other in some way. Each reaction can thus be represented by a system of Onsager's equations, coupled by weights (obviously different for each situation) determining the extent to which each partial function participates in the whole reaction. At the level of a subsystem, the coupling of the elements constituting it represents what we have called interactions, the principal feature of the brain's organization. In operative terms, the coupling (or interaction) is made possible only through reciprocal anatomical and functional connections, but also depends upon the amount of activity in different circuits and in different situations. The values of these coefficients thus cannot be fixed, because of the variability of nervous activity, even if the anatomical connections do provide for mutual influence on a stable basis.

In every physicochemical system the Onsager equations are subjected to a number of constraints of most general significance. The same constraints must therefore be respected when the equations are applied within the nervous system. A first condition is that the coefficients must not depend upon the forces, which is self-explanatory, otherwise the fluxes would always necessarily be defined in a unique way by the forces themselves. While this matter is easily settled in a physico-chemical system, it does not seem as easy in the nervous system, on account of the particular nature of the forces and fluxes (see the next section). Another, less self-explanatory, constraint is represented by the so-called Curie-Prigogine principle: in strictly isotropic systems and in conditions of equilibrium the coupling of forces and fluxes of a different tensorial order is not allowed. This would, after all, seem unlikely in the brain, since we have previously excluded the existence of perfectly isotropic, and true equilibrium, conditions.

In the nervous system we can approach homogeneity only on functional grounds and obtain only steady state conditions of work, so that the Curie-Prigogine principle should not bother us much. In the brain, however, situations of "symmetrical" activity conditions can be recognized, which endow the nervous system (or some of its subsystems) with nearly isotropic properties, a kind of true homogeneity (and in that case the constraint should be respected). We have to consider that in an isotropic system at equilibrium the relationships between the parts are symmetrical. Conversely, whenever a symmetry is recognized, through a symmetry of relationships between the parts of the system, we may presume the existence of a nearly isotropic situation and of equilibrium (or, at least, a steady state). As we shall see in later chapters, a symmetry sustained by widespread cross-connections between the parts may be likely, i. e. a symmetry of signal exchanges as a consequence of functional relationships maintained at a steady state level. Furthermore, as in a closed chain of chemical reactions, in the long feedback paths involving serially large numbers of subsystems, the fluxes (whatever they may be in the central nervous system) follow a closed loop. In the latter, at equilibrium, the fluxes need not vanish, the conditions of equilibrium being satisfied when the fluxes become equal (and thus constant), i. e. a true steady state situation. Now, whenever the fluxes become equal (a steady state situation) the coupling coefficients between any two fluxes and forces also become equal in the two directions, which is to say that the matrix of the coefficients is symmetrical. It should not be too difficult to create conditions of equilibrium steady states, and symmetry in brain activity. Actually, it would seem that symmetry of relationships ought to result whenever the system is left to itself, free to reach a stationary state, unhampered by exogenous forces.

### 6.3.1 The Problems of Entropy and Forces

The Onsager equations are subject to another constraint. In the case of paired equality of the coupling coefficients, the forces and the fluxes must be chosen so that their product equals "entropy". We have encountered entropy before as a measure of the irreversible changes of a system, due to the dissipation of energy in overcoming resistance, in heat and radiation, as well as to loss of information content (degradation of anatomical structure). Taking Ohm's law again as a typical phenomenological equation, the product of flux of electric charges and electromotive force equals the power dissipated in the circuit resistance.

Entropy is not yet uniquely defined in the brain, and different definitions seem to produce discordant results. In the CNS, perhaps because of the wide popularity of information theory among neurophysiologists in the last three decades, entropy has been measured (following Khintchin) mainly as the entropy of a distribution (the distribution of a time series of nerve impulses) which does not amount to much more than measuring the spread of a distribution, a simple, statistical descriptive method. If we are concerned with forces and fluxes, as in the preceding section, near equilibrium, in a steady state, or under symmetrical conditions in which the coefficients are equal, then the need for a physically meaningful measure of entropy imposes a precise constraint on the choice of the nervous forces and fluxes. At first sight, in the nervous system both the forces and the fluxes would seem to appear as nerve impulses (or at least strictly bound to them), so that what represents a force driving its conjugated flux of impulses is actually itself a flux of impulses conjugated to forces acting at the preceding synapses. Thermodynamics defines the forces as gradients of the state parameters (pressure, concentration, temperature) driving the corresponding fluxes. In the nervous system, measures of activity, even if performed on sequences of nerve impulses cannot "sic et simpliciter" be defined as state parameters (see, however, Chap. 8) at least in themselves, i. e. as they are recorded and somehow measured along the nerve fibers connecting the nerve centers. Actually, the focal point is at the synapses, whereupon the impulses impinge and from where the mechanisms of impulse generation are activated, i. e. where the appropriate forces and fluxes appear as thermodynamical factors. By sticking to the nerve impulses as such, one cannot see how the product of nerve impulses, constituting a force and a flux at the same time, could acquire the significance of a dissipation, since there seems to be nothing intrinsic in them akin to the concept of energy and of its dissipation, which is essential to the thermodynamical concept. Instead, at the synapses there are forces with true thermodynamical connotation, such as the concentration gradients of the synaptic mediators, and the chemical affinities with the receptors of the postsynaptic membrane. However, since each nerve impulse is assumed to liberate mediator molecules in definite amounts (quantal liberation), the concentration of the mediator would be a function of the nerve impulses arriving at the synapses. Unfortunately, it is utterly impossible to know the exact number of impulses impinging upon a neuron, because of the large number of branches from each nerve fiber and the presence of fibers from different origins on every neuron. Moreover, there is a different content of synaptic mediator molecules in each tiny nerve ending. It is possible to measure the concentration of the mediator with modern, refined methods of neurochemical identification, and use it, in turn, to find experimentally an approximate equivalent input nerve impulse discharge necessary to produce a similar concentration. A function of the output impulse discharge might well be chosen to represent the conjugated flux. At the synaptic level, accordingly, at which there are true thermodynamical forces, we can identify the corresponding dissipative processes in the synaptic "resistance" (measured through the Onsager relationships as the reciprocal of the "L" coefficient). Unlike physicochemical systems, in which it is generally possible to predict the amount of dissipation (increase of internal energy, etc.) the sources of dissipation being known and fixed, at the synapse level the coefficients of the Onsager equations are not constant. They represent the difficulty encountered by the mediator molecules in depolarizing the membrane to the threshold of impulse firing, but this difficulty is not uniquely defined. It is actually determined by the concentration of an inhibitory mediator, on one hand, and on the other by the phase of the excitatory cycle of the cell membrane following the generation of the preceding impulse, (as well as, probably, hitherto undefined, cytoplasmic metabolic processes). All these factors, may independently determine an infinite resistance, which is, in turn, the actual determinant of the silent periods in the impulse discharge patterns issued by a neuron. The coefficients, then, can partly depend upon state parameters (for instance, the resting membrane potential or a function of it) and can be partly time-dependent (inasmuch as they partly depend upon the phase of the excitability cycle) but not dependent upon the concentration gradient of the excitatory mediator, which is the force, in thermodynamic terms.

The problem of the forces in the central nervous system is by no means exhausted in these few lines, and will be resumed in Chap. 8, being strictly related to the picture of the state transformations, and together with the latter will be better focused as we progress.

The statistical approach has given us a first glimpse of the more general ways in which the brain operates. Through statistical mechanics it has been possible to obtain a theoretical definition of brain states as surfaces of constant energy covered by the path followed by the nervous system in its natural movement, that is, in a steady state. An extended and relaxed Gibbs principle – constancy of density in phase – teaches us that the ensemble of similar brain subsystems in their natural movements is expected to maintain constancy of relationships. Thermodynamics adds to the statistical picture of the time development of steady state brain activity a possible definition of the forces and coefficients at the synaptic level. However, this approach would obviously not lead us much farther. It could actually lead to the solution of partial problems, as in analytical research in which circumscribed subsystems are investigated. Yet, the phenomenological equations of Onsager have the simplicity and symmetry of physical laws, and can be used to build systems of equations representing in a simple way the state and behavior of a system, however complex. In the remaining chapters, where further approaches will be considered, we shall see the affinity of the matrix of the Onsager coefficients, on one hand, with the matrices of the correlation coefficients and of the components of the metric tensor, on the other, which we shall introduce as representatives of the nervous system's function.

# 7. States of the Brain

## 7.1 States and Stationarity

In the preceding chapter, statistical mechanics has led us to a theoretical definition of brain states and of their time evolution, when left free to follow their natural course (practically, under steady state conditions). We had there three alternative uses of the term state to mean one of the physiologically defined situations (such as sleep, arousal and many others), or the very fact that at any moment the nervous system is doing something (even if not something precisely physiologically definable), or, finally, the way in which nervous activity reveals itself, either as a stationary or as a time-dependent phenomenon, i. e. as a state of motion of some sort. Albeit the term state can also be used to refer to a single neuron, we shall use it only for dealing with the system, or a subsystem made up of a large number of components, as a whole. We shall try, therefore, to account for the multitude of successive aspects of brain activity in terms of the evolution of states, each representing the whole of the system at a particular moment. Each state may be regarded as constituting some kind of frame (or reference system) in which myriads of events can take place.

The multidimensionality of our brain, the ten thousand millions of neurons comprising it, with an undefined number of degrees of freedom, render it comparable to a volume of gas, and yet very different on account of the neurons' differentiation (which cannot be said of fluid molecules). Thermodynamics identifies the state of a system by means of a limited number of parameters, such as volume, pressure, temperature, concentration, etc. Since the molecules are not discriminable from each other, all the measures are statistical. Moreover, the definition of state is made at equilibrium, or in a steady state, (as we have seen, we cannot talk of equilibrium in the brain, only of steady states). However, we have not yet provided a clear definition of the parameters through which it might be possible to determine whether the brain is in a steady state. In physicochemical systems, the changes of volume, pressure, temperature, and reaction velocities are easily measured and the a steady state is known to occur in terms of the time derivatives of the measures. One first has to find out the nature of the characteristic parameters through which the brain state can be identified, still not an easy problem (see Chap. 4), since it amounts to defining for each system the forces, the velocities, and the related measures.

Another question concerns the time extent of a steady state. Non-living systems will remain stationary until the external parameters are changed. In other words, it is sufficient to maintain invariance of the relevant conditions. There is actually no indication whether it is at all possible to maintain a steady state within the nervous system by simply keeping the external parameters constant. (The answer is most probably no, since we know from experience how much the brain's activity may be independent of the surroundings.) When we do not know the conditions or exact mechanisms forcing the transition of a system from one steady state to another, we cannot possibly have any indication of the "natural" duration of a steady state (supposing we recognize one). This consideration is apt to cast some doubt on the tacit assumption we have made about the stationary nature of the brain's activity. Actually, talking of a steady state of the brain as a whole might be altogether unrealistic; the brain is made up of a large number of subsystems, each in turn made of a large number of classes of neurons (and each class, defined as homogeneous in functional terms, may include an undefined number of neurons). It is reasonable to suppose that the myriads of events occurring during any state might be associated with a steady state of some of the subsystems (or classes) but not of others; thus, the stationarity may shift from some subsystems and classes to others, while the events are developing. This observation leads to the conclusion that what we call the state of the brain can be represented only by a function of the state of all its subsystems, and the assumption that such a phenomenon as the steady state of the whole brain exists has yet to be proved. We can easily understand how the conclusion may have far-reaching consequences on what one ought to expect from the measurement of different brain parameters (such as, for instance, the activity of its many structures or classes of neurons). The essence of the question is, as we shall see in this and the following chapters, the "time" parameter in relation to neural activity.

## 7.2 Identification of Brain States

Nonetheless, the nervous system as a whole is continually in some "state" (a function of the state of its subsystems), which may be similar to or different from the preceding and the following states. Since what we call a state of the brain generally has a counterpart in some outward manifestations, i. e. behavior, there is a large number of non-neural parameters through which a state can be identified, such as blood pressure, respiration, muscular contractions (of limbs and viscera), eye movements, vocalizations, secretion of different glands, etc. Every one of these can exhibit some correlation with nervous parameters, such as frequency of brain waves or their patterns, and in general with the activity of the different nervous structures, through which we may conclude that the state is one of sleep, of excitement or quietness, of mood and so on. The point is that less than twenty states can be identified in this way through all the known parameters, whereas the everchanging patterns of activity of the ten thousand millions of neurons, with their multiplicity of connections, while engaged in the myriads of events in the frame of each state, certainly cannot be squeezed into a score of categories. This raises the question of the identification of the nervous variables, the state of which can in turn define the state of the whole. Conceptually, the billions of molecules of a small volume of gas are much easier to treat, since by exchanging the position and momentum of each molecule, the state parameters of the gas are not affected. This is a kind of operation one cannot perform with neurons. We can only try to identify classes of neurons endowed with similar functional characteristics, in order to justify average measures as in thermodynamics; but, even if feasible, at most this could reduce the thousand millions of neurons to millions of classes, a clearly impractical result, and probably a delusion. Moreover, nobody can assure us that the class will always remain unchanged under altered conditions (of stimulation, or generally of state of the brain). As was seen in Chap. 4, the answer to this question is also probably no. It is a common experience of neurophysiologists that, when submitted to changed conditions, the neurons of a class split up and rearrange themselves into other classes. Probably, the number of absolutely "fixed" classes can be reduced to comparatively few, such as the function-specific motor and sensory neurons, but, even among the latter an absolute invariance in every brain state is far from being demonstrated. Further, if the classes are not an invariant feature of the subsystem's composition, then the subsystems also cannot exist as well-identified functional units, but simply as anatomical subdivisions, with only very general specification of their functional properties. Therefore, the brain cannot be considered to be made up of stable subsystems and, thus, its state cannot be a function of the state of stable subsystems, but rather, of all its variables.

Leaving this point aside for the moment, we are confronted with another conclusion. The extremely large number of variables, and therefore the possible occasional extreme variability of the individual components' behavior, forces us to admit that there is an almost infinite number of successive different states. That (unfortunately, but obviously) only a score of them can be painstakingly identified, is exclusively due to the fact that we are presently using too few parameters and variables to identify them. Each added variable will split the less-defined previous state into more-differentiated "substates", and so on. It can be conceded that the vast majority of these newly defined states would have no counterpart in outward behavior (although this should be expected, for instance, in the many differentiated situations of an aroused state) but we can foresee that many of them will correspond to subjectively identifiable "internal" states (of consciousness, mind, or mood).

However attractive the prospect may seem, in practice the task appears discouraging on account of the intrinsic difficulties involved. Nonetheless, let us try to see what can we make out of the situation. In the preceding chapter we assumed that states could be distinguished on the basis of their energy content, when the systems follow their natural paths on surfaces of constant energy. To each of these surfaces was thought to belong a physiologically identified state, while the myriads of the subsequent steps, which must be considered as arbitrarily selected successive moments in the path followed by the system, were called the "elementary" states. The latter, would be undistinguishable from one another, being differentiated only by the continuously changing microdistribution of activity among the populations of neurons. Their not being discriminable, however, can only be referred to the fact that they belong to the same surface of constant energy. Thus, inasmuch as they belong to one of these surfaces, they form a "continuum" of elementary states, whereas the set of the surfaces (each at a different level of energy) forms a set of "discrete" states. However, the same statistical concepts which produced this picture, state that the energy of a given surface is a phase average, and must therefore be considered constant only in the average, thus allowing an effective discrimination of the elementary states in terms of their individual energy content; it is only the physiological significance of the statistical fluctuations that can be questioned.

More important, still, is the possible occurrence of dissipative processes while the system is in its natural movement. Statistical thermodynamics is centered on the problem of dissipation, always present in a real physicochemical system. Dissipative processes would progressively deprive the system of some of its energy, finally resulting in the "spontaneous" (we cannot call it "natural" in the statistical mechanics sense, see Sect. 7.2.1) transition of the system from one surface of constant energy to another, that is, from one physiological state to another. In other words, the system would tend towards thermodynamical equilibrium if energy were not added to maintain it where it is. The problem will be resumed later (Chap. 8); I only want to note now that it hints at the possibility of a continuum in what, at first sight, looks like a set of discrete, physiologically characterized situations, or states.

### 7.2.1 Discrete Physiological States

Let us assume, for the time being, that the set of states to which we can attach a physiological label (sleep, etc.) is discrete, and start with what might be a possible description of the situation, i. e. with the statistical aspect. The discreteness of the physiological states reminds us of the discreteness of quantum states (with which they could share a mathematical treatment). Quantum physics has developed only in the last sixty years, and has required more specialized mathematical tools to deal with such problems as electron states, and absorption and emission of energy by particles. The decisive advance was represented by Heisenberg's theory which fostered the development of a poorly known mathematical procedure, matrix calculus. The solution of the problems consisted of finally obtaining matrices with the energy of the states, and with the probabilities of transition between the states. It would seem important to investigate the possibility that quantum principles may be relevant to our brain states. Naturally, since we are not yet ready to measure the energies of the brain states, nor do we know all the possible states (besides the few definable ones just described), the whole matter should be deferred until these preliminary problems have been solved. However, the beautiful logic of quantum treatment must not be forgotten, and the representation of the matrix of states and of transition probabilities, in its conceptual simplicity, could also be retained for the brain's states defined through measures different from those of energy at quantum levels. In our case, the probabilities of transition are partly determined by cause and effect relationships, either known or unknown, somewhat randomly, but the matrix would still highlight the most probable as well as the "forbidden" transitions. After all, the occurrence of a particular state is the result of the causes determining the transition to it from another state, and of the simultaneous absence, within some limited interval of time, of causes of transition to another state. Thus, as we shall see later, transition operating mechanisms, at one hand, and states, on the other, though conceptually separable categories, are not actually separable, since neither is a stable process. In the next chapter we shall go deeper into this matter.

In quantum mechanics, transitions are either "spontaneous", that is, they are operated by fluctuations of the energy possessed by the system, or are "stimulated", that is, the jump from one state to another is produced by adding (or subtracting) energy. The same could be said of the nervous system, to which we can apply the identical reasoning and terminology. We know that our brain will change its state due to a driving force of some kind. In neurophysiology we use the same term, stimulation, because the transitions are generally induced by sensory stimulation or by direct electrical or chemical stimulation of key structures. But we also admit, although it is far from being proved, that there are transitions which are spontaneous, at least inasmuch as they are not produced by known stimuli. We could be tempted to conclude that spontaneous transitions are occasional events, determined only by chance, i. e. by statistical fluctuations of the values of a number of parameters while the system is developing its natural motion, because this would constitute a neat counterpart of the quantum mechanics situation found at subatomic levels. However, we should not let the subjective desire for symmetry lead us astray. It is much more likely that apparently spontaneous transitions are performed through the same brain mechanisms which are normally driven by an externally applied force. These would also be stimulated transitions, albeit the added energy would be supplied by sources not monitored by the observer (there are plenty of them within, as well as outside, the brain) and in barely adequate amounts.

All this would be right if we had the means of discriminating the states in terms of brain activity, as explicit functions of all (or a good many) brain variables. Unfortunately, this is not so. We have seen that we need to have recourse to variables outside the brain to help in the identification of even the few states we can think of. Again, the task would not be hopeless if we knew the exact relations (the correlation) between neural activity and those vegetative and somatic functions which ultimately depend upon the nervous system itself. However, until we are able to discriminate brain states through the brain's activity variables alone, there is a risk of encountering trivial results even with the more complex computations. On practical grounds, the problem concerns primarily clinical neurophysiologists, who deal professionally with human subjects, and must therefore rely only upon the data supplied by the electroencephalogram and a few other non-invasive methods of investigation. There have been a number of attempts, the more successful of which are those based on canonical analysis of the power spectrum of the EEG frequency bands, although using a limited number of variables. The results are extremely encouraging, since they demonstrate the actual possibility of obtaining discriminating functions in very good agreement with other discriminating methods. Still, the states which can generally be discriminated are only the three classical ones, arousal, rapid eye movement (REM) sleep, and non-REM sleep, far less than the (approximate) number of states, were the numbers of variables and of parameters much higher.

It is again apparent that the possibility of a finer discrimination of states depends upon the number of variables available. Let's see what can be done despite the obvious impossibility of obtaining measures from each of the thousand millions of neurons. The conclusion that the state of the brain is a function of all its variables does not necessarily imply that each variable needs to be individually monitored. Functions of groups of variables (or their classes) could suffice, provided the variations of the class composition, in functional terms, are somehow monitored. The method discussed at the end of Chap. 4, – the continuous correlation between integrated recordings of the activity of neuron pools – through which a functional change of an anatomically defined group in relation to any one of a manifold number of others can easily be detected, would seem sufficiently suited, at least to begin with, to monitoring the functional variations of the classes. The multifarious nerve events occurring in every state will each require a different constellation (or complexion) of functional groupings, resulting in different functional correlations between fixed anatomical groups. The method will give us this knowledge by detecting at each successive moment where, and to what extent, the activities of different neuron groups become more or less similar. The matrix of the cross-correlation coefficients, in which at each moment the elements may change in a continuum of values (though within a precisely defined range) could be taken to represent an instantaneous brain state, and their value at the time of transition to the succeeding matrix could represent a suitable basis for calculating the probabilities of transition from one to another complexion of correlations, i. e. from one state to another.

### 7.2.2 Redefinition of the Concept of State

Let us presume that notwithstanding the difficulties certain to be encountered we may be able to build a matrix of the states and a matrix of the transition probabilities, so that we have, though unavoidably incomplete, a picture of the situation. Our aim is to gain a deeper insight into the mechanisms determining a state and a transition. As a consequence of what was said before, what is necessary first is a redefinition of the concept of state. In a physicochemical system at equilibrium or a steady state, we can measure the stationary situation at any time, and at any moment we can consider the state as unchanging. But as far as the nervous system is concerned, the very term "state" is deceptive, since it conveys the idea of something stable as in a physicochemical system, allowing us to take measurements at any instant, as if we could "freeze" the state of the neurons at fixed intervals and still obtain at every "frozen" instant the same situation, definable through the constancy of the measures of a number of parameters. If we want to continue investigating the actual mechanisms underlying what we call brain states, this concept cannot be held any more, being highly unrealistic. Every situation we can think of, rage, mood, depression, excitement, sleep, thoughtfulness, etc., is not a frozen instant of our life, and thus not of our brain. Every physiological state is an actual, continuous processing of nerve impulses, continuously changing in amounts and patterns of activity in the different circuits, which do not permit the situation, as monitored through the activity, to be considered stable. Perception of sensory events, motor and vegetative reactions, and thought processes require everchanging patterns of brain activity, that we have no means of averaging in order to determine their eventual time-invariance (if the concept is ever applicable to such functions). Time itself, of which we may have a clear subjective notion (as we shall discuss at length later on) cannot be frozen, and brings with every increment the consciousness of something new, hence of steadily supervening changes. These are the reasons why no state of our brain can remain the same throughout, a fact we can perceive ourselves, if we only want to follow the course of those internal processes reaching the threshold of our awareness. What we may still call state, then remains only as a tag for situations grossly identified through outward manifestations, which, however, reveal upon detailed analysis an ensemble of motor acts, of vegetative functions, of psychic events produced by unceasing and everchanging internal processes, and betray for this very reason continuous, multiple and subtle differentiations of the brain activity at each successive moment, our personal way of living a state. I should say that we are not finding ourselves "in" a state, but rather, that our states are moving "through" us.

It is obvious that this way of considering a state, a situation in continuous evolution in its mechanisms, brings a profound change in what was said earlier about the discreteness of states. If what we have called a discrete state is not stable in itself, but represents a continuum of changing internal processes, the same must be true of what we have previously called "elementary" states. They can no longer be considered as a set of states of average constant measure (constant energy) as distinct from other sets belonging to different surfaces of constant energy (to other physiological states). Rather, as is also our subjective experience, all the elementary states constitute a continuous, micro-differentiated sequence, so that each physiological state abuts most necessarily on another. In other words, there would not be a situation in our brain in which the nervous activity actually moves on a surface of constant energy, simply because these surfaces could not exist, each elementary state (or infinitesimal step in the motion) being differentiated from the preceding and the following one.

The conclusion that there could not be a discreteness of physiological states certainly appears rather strange and conflicting with experience, since we "see" people asleep or awake for hours, then in what appears to us to be a definite physiological state. But, as we know already, in these situations there is a host of different stages, and I should like to repeat that more of the latter could be discovered by taking into account more variables and more parameters, which confer also to the outward manifestations of sleep or of wakefulness very different and variable characteristics. What I actually mean by saying that there is no discreteness of states is that the transition from one state to another is not an abrupt process, but a very much graded one, so that, for instance, the transition from sleep to arousal takes place through a continuum of intermediate steps, or intermediate states (perhaps, each with a particular duration of its own). Therefore, we say that one is fully asleep or fully awake only when one is a good many steps away from these intermediate situations. This means, also, that each physiological state gradually reaches an "acme", and afterward declines towards another state, which in turn will, eventually, reach its own acme or, returning, will climb again towards the former state's acme (as we see sleep do several times in the same night). Even in this crude picture of a situation, which might offer an actual multitude of branching paths for each of our system's states, it is clear that at several points and moments of the elementary states' evolution a selection of a particular direction, rather than of others, shall be needed. Here, we raise the question as to whether there are preferred and prohibited transitions, i. e. statistically most probable and highly improbable ones, which may account for the directions taken, or whether the choice of the direction in which the system evolves is always determined by the state of some particular subsystems. The answer given in the preceding section is decidedly in favour of the latter. Chance could not be allowed in the evolution of our brain states, (also see below).

Of course, this picture of the dynamics of states refers to the natural (not forced) development of events, to the apparently long-maintained states and their spontaneous evolution into other states. The process should go on unceasingly, if not disturbed, but I dare not foresee what, in the long run, must be the actual destiny of the system. A number of considerations prevent reasonable predictions. The abundance of feedback circuits in the brain is the basis of an ideal self-stabilizing system (as is generally believed to be the significance of the feedback paths). Somehow, self-stabilization might "trap" or "freeze" the brain in a given state, on a surface of constant energy (thus disproving my interpretation of the dynamics of states), from which it could be removed only by supplying some external force in order to wring it out and bring it to another steady state. This could again justify talk of discrete states; but, what would become of the actual natural movement of the system while the latter is trapped in a steady state is not easily predicted either. As time proceeds, the system should settle into an almost complete equilibrium (a situation hardly conceivable in a thermodynamical sense in the brain).

If we still want to consider the system's motion as being made up of a succession of instantaneous steps, or elementary states, what we may say is that these states ought to be degenerated (in a mechanical sense) and that nothing should prevent the situation from being prolonged indefinitely, though not in a physiologically definable way. On the other hand, it might be discussed whether things could not evolve in the reverse direction. The recirculation of nervous activity through the abundance of the feedback circuits present in every structure, instead of producing self-stabilization of the activity, and therefore indefinitely "freezing" the state in its parameters, might, on the contrary, supply the energy necessary to perform a state transition, by inducing those energy fluctuations on which spontaneous transitions (in quantum mechanics) are based. The positive feedback character of some of the reverberating circuits, and the inhibition of inhibitory structures, could induce these effects (and, of course, the transition would again be of the stimulated type). On such mechanics is based a still current interpretation of the sleep-arousal alternation.

Introducing into this picture external factors, such as sensory stimuli, changes of concentration of some body constituents, products of secretions, etc., induces changes in the situation. Whenever these factors operate concentrated in time, i. e. abruptly, the state changes suddenly, the transition manifesting itself with startling phenomena, such as sudden arousal from sleep with the motor and autonomic reactions of alarm and defense, according to the nature of the arousing stimuli. However, during the several seconds required for the transition to be completed, a fairly quick succession of intermediate stages develops, a fact not specific to any type of transition. There is experimental evidence that these external factors which promote state transitions are not necessarily contingent, but are steadily operating as part of the normal life of the brain. As a matter of fact, the brain is continuously processing sensory stimuli, even in deep sleep (moreover the stimuli are also not usually strong, but mild ones), has to maintain muscle tonus, reacts to changes of blood constituents, provides for the secretion of hormones, has rational processes, feels emotions, changes its memory content, etc. All this certainly represents a more realistic picture of the brain's ways, and leads us to conclude, once again, that there cannot be a "natural" movement in our brain, that the system is never moving by itself. (Would it otherwise quickly tend to equilibrium?) Rather, the changes which constitute the transition between elementary states are always driven, or to use the terminology of physiology and quantum mechanics, are always stimulated.

In a way, we have to admit that there is actually a discreteness of states, if not at the level of elementary states, in which the differences might be very small and as such allow the consideration of the set of these elementary states as a continuum, at least at the level of behavior, where clear-cut (that is, overtly discrete) differences are discriminable when the states are near their acme. However, I cannot refrain from stressing that discreteness must not be intended in quantitative terms (as determined, for instance, by energy or energy-related measurements) but that we have to maintain the concept of discreteness in qualitative terms. The latter can be appraised only through the matrix of relationships, i. e. the patterns, between all the variables, as proposed in Sect. 3.2. From the quantitative differences among the elements of the matrix stems the qualitative difference between the matrices, a fundamental argument for quantum states, as well as for brain states, (see the next section). The assumption of identity of the two situations, however, requires us to solve the problems of "energy" in the nervous system, otherwise there will always be insurmountable differences between quantum states (like those of hydrogen atoms) and physiological, behavioral and elementary states. While for each type of atom there are unique values for each energy level (each state, actually), in the nervous system the absolute values of the variables might well be similar in different states. Thus, to the same physiological state could be assigned different energy levels, indicating that a brain state would be independent of a merely quantitative definition (such as a phase average) of its parameters.

## 7.3 Quantum State Principles and the Brain

It is clear, from the preceding considerations, that there are several similarities, as well as some important differences, between the brain and a quantum mechanical system. Until now, however, the similarities have appeared to be more superficial than substantial. In quantum mechanics, the state of an atom may undergo either a spontaneous or a stimulated transition, but the result is always dominated by some fundamental uncertainty, so that it is aptly defined only in probabilistic terms through a particular set of statistics. In the nervous system, it is true, we have discussed the possibilities of spontaneous and of stimulated transitions. However, while from the results of the experiments of the last three decades there is no doubt about the stimulated ones, the occurrence of spontaneous changes has certainly not been sufficiently investigated. In the brain the transitions do not appear to be intrinsically uncertain, but are always accomplished through precise patterns of change in the activities of different brain structures (preceding, accompanying and following the transition), thus are represented by definite sequences of events. In every case, the transitions themselves follow ordered sequences in time and are performed through the mediation of well-known neural structures, the diffuse projecting systems, the main physiological function of which is that of controlling the level of alertness and, through that, of consciousness and performance of the brain. Partly, this is effected through the mass activity discussed previously (Chap. 5) and the consequent changes of the average level of impulse discharges in widespread areas of the brain, and partly also through the signals in the specific interneuronal connections. In quantum mechanics, an energy gradient (absorption of energy or spontaneous energy fluctuations) is the unique transition-inducing mechanism. In the nervous system, on the other hand, we still cannot define an event analogous to energy absorption. Yet, even conceded that somehow we may talk of absorption of energy, not being at quantum level the system may certainly absorb energy in an extremely graded way, so that the transitions, though stimulated, would still maintain a graded character in the complex of the system. (There is some experimental evidence for this.) This mechanism most probably accounts for apparently spontaneous transitions, since a system, if it could be really left to itself, i. e. without adding energy to it, would probably settle for some time on an energy-constant surface, from which it could be removed only through losses of energy by dissipative processes.

If we accept that transitions of physiological states are not to be attributed to random fluctuations of energy levels, there should be no reason for further pursuing the quantum state approach in order to analyze their mechanisms. However, an important suggestion comes again from quantum mechanics, notwithstanding the difference between the situations. A fundamental principle says that every state results from the linear superposition of two, or a number, of other states. Quantum theory attributes these different states to the same physical system (in the limit to the same physical object). Apparently, we could not apply this principle in the nervous system, since the brain is made up of a number of subsystems and classes (however unstable), each in some state of its own, and all being different "objects". The "superposition" of the states of each subsystem certainly defines in the brain the state of the whole, without necessarily being a linear function, such as the sum, of the states of the subsystems. A trivial observation, since in this case the quantum principle would not be referred to a single system, or object, but to an ensemble of different structures, and there should be no reason for applying any quantum analogy. Nonetheless, although the quantum principle refers to one single, indivisible system (for instance, an electron being possibly in two different states simultaneously) the same principle could be applied when a unique system is made of different parts or "properties", provided it exhibits an indivisibility on anatomical and functional grounds, at least for a limited interval of time. Certainly, stated in this way, what we have looks like a simple analogy (and not a very appropriate one either), but we may try to analyze it better. Each subsystem of the brain has been assumed to be made up of a number of classes, or groups of neurons, organized so as to build up that subsystem. It is clear that this subdivision could go on creating sub-subsystems indefinitely. The state of the last subdivision (class or group) creating a "unity", then, is determined by the relationships between its parts (as we have previously seen, by the degree of their reciprocal, functional correlation) since each of the parts can contribute to the state of the whole only in some measure proportional to its own state, and conditioned by its strict, functional relationships with the other parts. Formally, the quantum principle of the superposition of states says that the contribution of each state to the resultant state is represented by a coefficient, merely the "weight" of the contribution, its probability. In the nervous system too, we could assume the existence at some smaller unitary level of a principle of linear superposition, inasmuch as at that level the state is defined by the coefficients linked to the state of each of the elements comprising that particular unity (i. e. the coefficients of the correlations between each element and the others). The real point is that quantum states are represented by linear differential equations, which can be considered as the linear combinations of other linear differential equations. If we want to apply the same principle to the states of brain subsystems, of brain "functional unities", it ought to be demonstrated first that their states too can actually be represented by systems of linear equations. How this might eventually be the case, will be clearer after Chaps. 9 and 10.

There is a point, however, at which quantum mechanics and physiological systems shall never coincide. The principle of superposition of states in quantum theory is rooted in the assumed fundamental indeterminacy of the state of matter, which, most naturally, leads us to a statistical solution of the problem, the coefficients having only a probabilistic significance. On the contrary, the physiological principle of superposition of states within the brain does not necessarily have a probabilistic basis. It refers to different subsystems, and the state of each of them can be defined (as we shall see in Chap. 8) through dynamic parameters, and is determined by the action of the diffuse projecting structures. Therefore, the state of the whole system can result from the superposition of the states of its subsystems, but the coefficients defining the weight of each of them represent absolute measures, not necessarily probabilities.

Obviously, since the source of determination in the physiological states seems to reside in the state of the diffuse projecting system, it is justified to ask how the state of the latter is determined, because if it were fully dependent upon chance, or upon external, uncontrollable events, then, although its further effects on the other brain subsystems could be perfectly determined, the whole situation would suffer from an indeterminacy, as in the case of quantum states. That the situation is not in these terms is indicated by two orders of facts. The diffuse projecting systems are reciprocally connected with "lower" brain stem and spinal structures, on the one hand, and with "higher" brain structures, including the cerebral cortex, on the other, therefore they are part and parcel of feedbacks chains which leave little space for indeterminacy, since they probably tend to stabilize a state, and hence, to determine it more precisely. Furthermore, the changes of state of the whole system which are the most clear and constant (thus, not left to chance) in practically every species, are the cyclic alternations of rest (sleep) and activity (arousal) making up the so-called circadian rhythm, with the associated cycles of readiness. The cyclic alternation is not destroyed, but simply altered when the being is no longer subject to the pacemaker, or rhythm-organizing effects of the alternation of light and darkness, of hunger and satiation, thus revealing some intrinsic, independent regularity (even if not as precise as clockwork, at least as dependable as that of a relaxation oscillator) of the state-determining structures.

### 7.3.1 States and Vectors

The representation of the state of a system as the linear combination of other states is easily obtained in the vector domain, and one of the major achievements of quantum mechanics has been the development, with the theory, of the mathematical formalism to support it. Thus, the linear superposition of two or more states is best represented as the superposition (sum) of vectors, the resultant being a "ket" vector. The basic idea embedded in the mathematics of ket vectors, is that the parameter defining the state is not the length (the module) of the vectors, but their orientation in space, the direction (a function of their spatial relations). As a consequence, if the coefficients defining the contribution of the vectors to the resultant ket are in a given ratio to each other, the resultant ket will again be specified by the ratio of the same coefficients. Assuming the same quantum principle of the ket vectors and their law of superposition for the states of the nervous system we find the formal justification of the interpretation given in the previous section. The procedure is straightforward: each "unitary" brain subsystem, which is made up of parts, is specified not by the measured amount (the equivalent of the modulus) of the activities of its components, but by the ratios (that is, the reciprocal relations) between the activities. Of course, the state of the whole system (its ket vector) will be specified by the ratios of the coefficients associated with the vectors of the components. This way of defining the states of the nervous system sounds much more reasonable than the one based on energy content discussed earlier, because it seems fit to serve as a theoretical basis for the constancy of functions that the brain must preserve under widely varying conditions, generally involving widely varying levels activity (i. e. different lengths, or modules, of the component vectors).

The independence of a state from the modules of the component vectors, and its dependence upon their mutual relations (as defined by the ratio between the coefficients) bestows upon this representation of states a pictorial character, very useful for a biologist. However, it all rests on the representation of nervous system variables as vectors. In Chap. 4, we found that a common basis for quantification of nervous activity, suitable for direct comparisons between recordings of different variables, that is, for analyzing their reciprocal relations, was the time integration of the electrical signals (single cells or large populations). Using this, all our nervous system variables can be represented as vectors in a space with as many dimensions as there are variables. By computing the correlation between pairs of variables one reaches a double goal, normalizing the ket measures, thus having only the direction of the vector to represent the related variable, and determining the relationships between the variables, measured as the cosine of the angle formed by the respective ket vectors. In the nervous system, the state is better visualized through the pattern of relationships between the component subsystems, as represented by the matrix of the coefficients, than through the direction of a single ket, in which all the information about the state of each subsystem of a multitude would be lost. To be able to concentrate in one single ket vector the multiple aspects of a complex situation, such as any one of the brain states, like an arrow pointing (for each state) in a different direction within such a multidimensional space, is enticing our fancy, and may satisfy our longing for simplicity. In practice, unfortunately, we are still far away from the goal, while, on the other hand, we shall find the matrix representation to be extremely useful, being aligned with the other approaches to be discussed next.

Before closing the chapter, it seems necessary to take heed of the gradual modification of our way of looking at the problem of state representations in the nervous system. The overall picture has changed: we have shifted away from an ensemble of subsystems (in Gibbs's terms) moving in space in their natural motion and maintaining a constancy of reciprocal relations, and we look now at a multidimensional space bounded by axes, the ket vectors of the state of each subsystem, not immobile but continuously changing their orientation relative to each other, so that the representation of that space is continuously changing its shape. Somehow, this also seems to change the perspective of our endeavour. When, at each instant, each of these vectors exhibits a different orientation, introducing some change in the volume of space bounded by them, we have to accept the idea, or the picture, of a volume which continuously transforms itself through a minuteness of changes (in which, perhaps, it could be arbitrary to recognize a series of even minute steps). In practice, since we are forced to introduce arbitrary time constants for our measures, the transformations of the whole system would be treated as a discrete, but continuous series of steps. We do not yet know whether there are really "characteristic times" in the nervous system's functions, that is, whether characteristic durations of the different neural events do exist (besides the basic event, the nerve impulse). This forces us to high-rate (short-time) resolution in the computation of the relationships between the kets which should define a state, lest we miss relevant information. Only after collecting enough experience shall we be able to "adjust" the resolution rate (the time) in our computations to the needs of the situation.

While the problems connected with the dynamics of continua will be treated later on, we shall limit ourselves here to identifying immediate goals. Since the number of variables is extremely high, we could attempt, first, to sort out the common features of the subsystems, in order to facilitate the computation. This is a goal to be typically approached by statistical methods (such as, for instance, multifactorial analysis). However, in the true spirit of a neurophysiological investigation, rather than identifying states, we ought to concentrate our efforts upon the identification of the mechanisms of the changes, that is, upon the laws of transformation determining the transitions and the continuous evolution of our brain's states. To this aim is devoted the next chapter.

# 8. Dynamic Laws and Transformation of Brain States

## 8.1 Statistical Laws and Deterministic Laws

Our conclusions about the states of the brain and their quick succession in time have led us to consider the dynamics of the mechanisms operating the changes. As thermodynamics prescribes, state transitions are determined by the exchange of energy, and although we are still not at ease with energy measurements, we must consider the forces involved in the transformations. Hence, a change of our frame of thought now seems to be necessary. Of course, statistical thermodynamics also deals with forces, with potentials as well as their fields, but the concepts are borrowed from other domains of physics and do not need to be based on statistics.

So, we have finally reached a point where we have to abandon the prevailing statistical reasoning, so helpful in many instances, and favour an approach based on more deterministic principles, such as those of mechanics. Statistics induces one to recognize a primitive, fundamental randomness in brain functioning. In the same way, dynamics might lead one to accept the idea of a rigid determination governing the brain's behavior. In the latter case, however, whatever rigid determinism might govern the behavior of the individual components (single neurons) would be completely obscured by the behavior of the whole in which many differently determined individuals are participating. (This is actually the true basis for applying statistical methods to the analysis of the properties of the nervous system.)

Moreover, uncertainty might sneak into the behavior of the system, even if regulated by strict deterministic laws introduced by (hitherto undefined) dissipative processes which would prevent, as we have seen, not only complete reversibility of phenomena, (which would not be assumed anyway in the nervous system), but principally, full development of the effects of the applied force. Other sources of uncertainty are the methods of quantification, i. e. our measurements. It was said in Chap. 4 that the information contained in the EEG is not expressed in the individual waves, but in their time sequences, in a complex of waves. This led us to the time integration of the recorded variables, and the computation of their power spectra. Both integrated voltages and power spectra are only averages of different properties of the same electrical signals issued by the neurons, which perform functions with a (presumably) characteristic time for each function (the two procedures being connected through a Fourier transformation). Thus, again, averages must be computed, even if in this case we are not following statistical reasoning. A type of average is necessary on account of the principle that the state of the system cannot be identified in a frozen instant, but in parcelled, characteristic time intervals of a continuously evolving situation. In this way, measurements of the electrical counterpart of the activity, linked to that situation or state, are the time-quantified variables on which the computation of the reciprocal relationships between brain structures is to be based. This statement, then, substitutes for the (altogether arbitrary) assumption that the only meaningful measures in the nervous system are time averages, because the time series representing the brain's activity would be of a stochastic nature.

It is a fundamental point that nervous activity can have a meaning only if viewed within some characteristic time interval. The next three chapters are extensively concerned, directly or indirectly, with the (mainly unsolved) problems posed by time as one of the principal factors in brain function. It is actually most probable that each situation would require time periods of a different averaging length (i. e. a different parcelling out of the time intervals) depending upon the level of detail of the information in the time sequences, or, in other words, upon the rapidity of the changes. It ensues that only by chance can the length of the time intervals selected for our measurements be the right one for describing the phenomena. An automatic, time-varying strategy should be invented (as I was saying at the end of the preceding chapter). There is also something more in this. Time defines the sequence from before to after, in which events develop, i. e. in which the agents of development, forces, exert their action. Here, the concept of force as the origin of the changes enters our discussion.

As was said in Sect. 6.1.1, however, the concept of cause and effect relationships in deterministic laws, like those of motion, does not prescribe a unique and defined temporal direction, since the laws by themselves are symmetrical. An identical situation is found in the brain, because the causes and the effects are, in the very end, the activities of different structures, interconnected more or less directly through circuits with many feedback pathways. Thus, in a circuit in which the nerve impulse trains can return, albeit modified in pattern and number, to some point through feedback pathways, the distinction between cause and effect becomes futile. The only possible distinction is between what comes first and what comes later at some given point, when known forces start operating at known times. The overwhelming emphasis attributed to the nerve impulses as carriers of information has prevented us from regarding nervous activity as the seat and origin of forces, whereas the latter are part and parcel of each nervous system's operation, because only through them can the development of events be explained.

## 8.2 Dynamic and Nervous Parameters

We shall try now to identify the forces which act within the brain, and to describe their operation. There is obviously an initial difficulty to overcome, if we want to treat nervous activity in the same terms as a mechanical system, i. e. to identify the counterpart of the dynamic parameters in nervous activity. The representation of motion, within the brain, as the displacement in space of the nerve impulses from one neuron to another could correspond to a visualisation of the impulse trains as carriers of energy (mechanical or electrical) exerting a force at their terminals. This representation, however, would be deceitful in two ways. As we have already seen, the nerve impulses are not the only manifestation of nervous activity, which also includes the graded membrane potentials preceding and following the impulses, as well as other potential oscillations, seemingly unrelated to impulses, and none of these is a "travelling" activity. The nerve impulses themselves are not actually moving in space, but are generated anew at every successive point along the nerve fibers connecting the neurons, and only by extending the concept can we discuss the velocity of conduction of the impulses along a pathway. Therefore, the force that the impulses may exert at the synapses cannot be likened to that of a bullet or an electric current. The movement of non moving parts, the neurons, can only be endowed with the old philosophical connotation of a change in time (see later in this chapter).

Dynamics identifies two parameters of motion, mass and velocity, and derives from their product a third parameter, momentum. Mass is a term we have alreay encountered and discussed at length in Chap. 5. The property of mass activity in the brain was compared with that of a gravitating mass, of a body which exerts its influence in a measure proportional to the sum of the influences of its components. As a consequence, nervous mass activity can be identified only by some summable function of its elementary components (in a functional sense), either the individual postsynaptic potentials, when the phenomenon is viewed at the level of the cell membrane, or the nerve impulses, when considering the activity of the nerve fibers which connect the neurons. However, as we said previously, the nerve impulses cannot be looked at as material objects having a "rest" mass, not being projectiles or moving in space in any way, so that they can be represented only through their rate of appearance at the synapses, the only place where an action can be exerted. As a matter of fact, the impulses are much more similar to "massless" particles (like photons and neutrinos), and the situation in the nervous system could be treated accordingly. If we assume dn/dt (the impulse rate) to represent the frequency $\nu$, then the impulse action at the synapse could be represented as $h\nu$ (where $h$ is Planck's constant, the quantum of action). In this case, as for photons, the momentum would be represented by the ratio of the energy $h\nu$ to $c$ (the velocity of propagation of light in a vacuum). The concept of the impulse discharge, as represented through $h\nu$, will be resumed later on, in Sect. 12.3.2.

Does that represent the answer to our problem? Actually, it would seem that the matter has been settled rather easily, but it is really not so simple. We have introduced definitions using the constants $h$ and $c$ about which we know nothing as far as the nervous system is concerned. Planck's constant, with the dimensions of energy $\times$ time, could then be assumed to be universally valid in the nervous system as well, representing there the quantum of action – associated with the impulse – as in every other physical system. (However, as hinted previously, the essence of the nerve impulses is not mechanical or electrical, and moreover, the time involved in the constant could even be assumed to be some characteristic time interval related to the state, which would lead, in turn, to a relativistic transformation of $h$.)

We will not proceed further with these wild speculations, which border upon physical heresy, and for which there is so far no evidence of a real need. As for the constant $c$, it is justified as follows: the concept of $c$ as a limiting velocity, rather than as an electromagnetic unit, would seem more suitable for the nature of motion within the nervous system, but a discussion of its properties will have to wait until much later (Sects. 12.3 and 12.3.1) since it is strictly concerned with some relativistic aspects of nervous functions. On the whole, it seems that rather than with a classical mechanical system, we are again confronted with a quantum mechanical system, and we should try to unravel its intrinsic properties according to quantum principles. Before we do this, we still have to further explore (as in Sect. 6.2.1) the possibility of a classical approach, through the formalism of generalized coordinates.

### 8.2.1 Generalized Coordinates and Energy

Leaving aside, for the time being, the problem of the nervous dynamic variables, let us concentrate on the nature of motion within the nervous system. Dynamics identifies two fundamental types of movement, one having a constant, and the other a time-varying velocity, respectively a "uniform" or "inertial" motion, and an "accelerated" motion. Moreover, classical dynamics also states that a body conserves its motion when it is not subjected to forces, i. e. the dynamic laws are conservative. We have already discussed the matter of uniform motion in Sect. 6.2.2, in which we dealt with the ongoing activity of nerve cells. There, we concluded that the neurons have no mechanism of self-excitation, and that the generation of impulses is always a driven process, bound to the impulses in the input synapses, time-dependent, inasmuch as it is damped by a time-varying membrane resistance and by inhibitory influences. We have tagged this type of activity "quasi-uniform" motion, since it is sustained by accelerating forces, each short-lived, but maintaining, on the whole, a steady state situation. All experimental evidence is consistent with the fact that in no neuron is the activity constant for longer than a few seconds at most, if not submitted to a steady bombardment of afferent impulses. The steady state situation thus created seems to be the most we can obtain in terms of constancy of activity, the situation closest to truly uniform motion.

Now, if we accept, at least tentatively, quasi-uniform motion of neurons as a substitute for physical uniform motion, with conservation of energy, then we are confronted, once again, with the necessity for measuring energy in the moving nervous system. As we have seen, the problem posed by the existence of systems with many degrees of freedom and non-moving parts, such as our nervous system, was solved by Lagrange with the formalism of generalized coordinates. According to Newton's law which defines motion through position and velocity parameters, Lagrange indicated with the "generalized position $q$" any parameter which defines a state of the system, and as a "generalized velocity" (since there could be no moving masses) the first time derivative of the generalized position, $\dot{q}$ (see Sect. 6.2.1). The Lagrangian function $L$ is a partial second order differential with respect to $q$ and $\dot{q}$, and is equal at any time to the difference between the kinetic energy $T$ and the potential energy $V$ of the system, the so-called "kinetic potential". (The Lagrangian tends to an extreme, i. e. the difference $T-V$ tends to become constant.) The disadvantages of Lagrange's method are mainly that the number of the second order equations quickly becomes impracticable as the degrees of freedom increase. The advantages are the beautiful simplicity of the formulation, allowing representation of complex composite systems and the solution of the equation of motion, which in the computer era may imply a bearable length of time. It is clear that Lagrange's method is theoretically suitable for ongoing brain activity – accepted as representative of uniform motion – in which energy is conserved. (The kernel of the Lagrangian method is the deterministic representation of a conservative system, although the method allows for the introduction of corrections required for non-conservative systems, those with dissipation of energy.)

A somewhat different, and for our purposes, perhaps more useful approach, still based on generalized coordinates, is that of Hamilton. The Hamiltonian function (see Sect. 6.2.1), $H = T + V$, is equal to the total energy of the system. Hamilton also gives an explicit definition of the momentum $p$ as a partial differential of the Hamiltonian and $q$. The partial differential equations of Hamilton, determining the state of the system, are the "canonical" (fundamental) equations. Their principal advantage (besides simplifying the computation involving first order differentials) is that they remain invariant under many transformations (such as the transformation of coordinates), i.e. any law expressed in canonical form remains invariant in different reference systems. The latter property is important for our theory of brain function, since, as we shall see, in general it is not necessary to solve these equations directly, but to use them to demonstrate that the brain maintains invariance of function when one reference system is transformed into another (see later in this chapter). For our immediate purposes, the Lagrangian, and particularly the Hamiltonian, are important functions because they fulfil our expectations of a method for defining the energy of a system without necessitating a recourse to statistics. Moreover, through the gradients of the energy we have a way of also defining the forces in the brain, completely independent, in principle, of statistical and thermodynamic measures, which was our aim when we set out to find the mechanisms of changes of state in the nervous system. We should not lose sight, however, of the fact that the velocity of our quasi-uniform motion is constant only in the average, and therefore, that we shall never be completely free from some statistical uncertainty.

## 8.3 The Dynamic Forces in the Nervous System

Having found an approximate definition of uniform motion, now we can examine how the nervous forces in the brain transform a uniform motion into an accelerated one. Forces are defined in thermodynamics as gradients of pressure, of chemical potential, chemical affinity, etc. In dynamics forces result from gradients of mechanical energy (see the Lagrangian and the Hamiltonian functions) and are measured through the effects produced, such as acceleration of a mass or change of momentum. The approach could be sufficient for a number of purposes, but is by no means unique nor sufficient for defining the force functions within the nervous system. Moreover, there is another preliminary question: why should there be gradients of energy in the nervous system, in which the richness of connections between the neurons would induce us to expect, much more reasonably, a fairly uniform energy distribution throughout? Actually, this situation of near-homogeneity of distribution of the energy would be attained, most likely, in the case of steady state activity, and a gradient there could be created only by the addition of energy concentrated at some point, thus originating a force which disturbs the stationary quality of the system.

Also in Lagrange's and Hamilton's formulations are the two forms of energy, potential and kinetic, kept distinct; it is worthwhile to inquire whether the distinction also makes sense for the nervous system. At first sight, one might assume that, as in a pendulum, there is a position and a moment, i. e. a situation, in which potential energy is at a maximum and the kinetic energy is nil. In the nervous system every activity, consisting either of impulses or of brain waves, was made equal to a motion (thus, energy in effect) and the impulses, as such, cannot be conceived as having a rest mass, so their energy can be represented only by the product of their rate (one could also consider the brain waves and their frequency) and of Planck's constant $h$. Therefore, we can identify a potential energy only where and when there are no nerve impulses or the strictly associated membrane potentials preceding them. That is, at the place of the physical connection between the force and the effects of the force. The unique place in the nervous tissue where this occurs is at the synapses, where the transmitter molecules set up the actual mediating process between the nerve impulses at the input of a neuron and the impulses at the output. It seems reasonable, then, to recognize the potential energy as represented by the number (or the concentration) of mediator molecules present in the synaptic vesicles of the nerve endings, an energy which becomes actual only upon the arrival of nerve impulses. The latter, liberating molecules of transmitter (apparently in quantal amounts for each nerve impulse), create a space gradient of concentration which constitutes the force acting on the postsynaptic membrane. The effect is measured through the postsynaptic potentials, upon which depend, finally, the successive generation of the output nerve impulses. It is easy to see, also, that the synapses are the only places where interference between different forces becomes possible.

The interesting feature of this chain of events is that when the synaptic vesicles liberate the mediator molecules, a process is immediately initiated which synthesizes them anew, i. e. a process restoring the potential energy. It is debatable whether it is really the concentration gradient which constitutes the force, or another aspect of the molecular activity, such as the affinity of a chemical reaction, since the synaptic effect is conditional upon the binding of the transmitter molecules with the receptors on the post-synaptic membrane. By originating the post-synaptic effects, membrane potentials which may or may not be followed by new nerve impulses, the potential energy is transformed into energy of action (kinetic energy). The overall process of synaptic excitation takes on the appearance of an aperiodic oscillating system, with a driving force, the input impulses, and a restoring force, the metabolism of the mediator molecules. The aperiodic character is imposed by the asynchronous arrival of the excitatory impulses at different nerve endings.

Like every mechanism based on chemical reactions, the generation of nerve impulses by a neuron has an intrinsic indeterminacy, which can be measured as the probability of transmission, and should be overcome, at any rate, by a "safety" factor. The latter is represented by the ratio of the number of transmitter molecules liberated, to the threshold amount for membrane excitation. However, while the safety factor along a nerve fiber (which is expressed as a function of the density of the electric current crossing the membrane) ensures that the nerve impulses can be steadily generated all along the fiber, it is doubtful that there is a safety factor which is always effective at the level of the synapse, since this property would make the synapses nothing but so many relays. Of course, if the synaptic junctions were to act simply as relays, there would be no reason at all for their existence, on account of the fact that the nerve impulses are not actually travelling along nervous circuits, but are newly generated at every place in the circuit, (so that there would be no need to generate them at intermediate synapses). The presence of a synapse makes sense, inasmuch as it represents the only place where it is possible to introduce changes in the impulse trains, or rather to shape the impulse trains through the interaction of forces exerted by different input circuits. A successful synaptic excitation through one of the circuits is affected by the simultaneous presence of two opposing factors: the impulses at the input side in sufficient number (since they need to be summated, either in space at nearby synapses, and in time at the same synapse), and by inhibitory influences. The problem is not a minor one. The forces necessary to accelerate a system must be higher, the greater the momentum of the system's motion. The synapses constitute the place at which the forces act to change the momentum of the post-synaptic neuron's motion. Therefore, we may expect three different situations. If the ongoing activity of a neuron (its actual motion) has a high energy, and thus a large momentum, it could be effectively acted upon only by an added, strong driving force, though in the extreme case the whole neuron's capacity for producing impulses could already be used up by the ongoing drive, and the added force would remain ineffective (we should also take into account, in this connection, the gating property discussed in Chap. 5). Thus, the system's motion does have an actual limiting value as if its momentum were infinite, a consideration to keep in mind for later on. On the other extreme, in the absence of any ongoing activity, zero momentum, an added drive would easily modify the membrane potential, but in the limit the whole available force could be used up in decreasing the membrane potential to the threshold of impulse firing, and it might as a result become insufficient to generate nerve impulses.

The need to overcome a threshold of membrane depolarization in the nervous system corresponds to the need to overcome inertial forces in dynamics when a mass has to change from a state of rest to one of motion. A more favourable situation is the presence of a moderate level of ongoing activity, since an added driving force which is also moderately strong should be facilitated in its effects by the presence of a sufficient subliminary excitatory state, so that more nerve impulses can be produced, thus increasing the level of the ongoing activity. There are no doubts about the existence of subsystems maintaining a central excitatory state, either directly or indirectly, in almost every subsystem of the brain. As a consequence, the central excitatory state would not be allowed to undergo "chance" oscillations, being actually enhanced or lowered only according to the type of physiological situation (such as sleep, arousal or other different states of consciousness). One might say that the synaptic "machine", when driven at a level of moderate kinetic energy (a moderate ongoing activity), a quasi-uniform motion, is like the engine of a car purposefully kept at a moderate pace, whence it can be easily accelerated by adding small amounts of driving force.

The forces we have been able to outline so far are of two types. There are true thermodynamic forces at the synapses (concentration gradients or chemical affinities), and mechanical forces represented by the gradients of energy defined by the Lagrangian or the Hamiltonian functions, once the generalized coordinates of position and momentum have been defined. It is clear that this last point is at the core of our problem. Before going into it, however, we must recall that also the synaptic force (the thermodynamic force) could be expressed, in principle, in terms other than concentrations and affinities. Transmitter molecules are liberated from the synaptic vesicles in proportion to the number of nerve impulses impinging upon the synapse, so their concentration is always a function of the nerve impulses. As a consequence, the force would also be representable as a function of the nerve impulses, i. e. in terms of their number or rate, as an "impulse equivalent" force. Once the relationship is found, it may allow the unification of the terminology of all the forces acting within the nervous system.

### 8.3.1 Nervous Generalized Coordinates

We ought to be very careful in our choice of the nervous counterparts of the generalized positions $q$ as representatives of the state of the system, since through them we should attain a definition of $H$, $L$, $T$ and $V$ compatible with the definition of kinetic and potential energies in terms of nerve impulses. Since the $q$ positions should contribute to the definition of the state of any subsystem, or class or group (or even of individual neurons), the number of the $q$ positions necessary to define a state depends upon the degree of detail we want to achieve in the description of our system. As for single neurons, one of the most representative functions of state is either the transmembrane potential, or some function of the latter, such as the instantaneous rate of impulse discharge (the set of all the simultaneous values for the entire neural population would define an elementary state of the whole). In order to unify the treatment of energy and force in the same terms, we may well nominate the instantaneous impulse frequency as $q$. It must be stressed, obviously, that what we mean by state here indicates only what a structure (in the extreme case a single neuron) is doing at that moment, and is not the state of the system in terms of function performed, which must be viewed within a characteristic period of time for each function. However, it is our final goal that $q$ might be so selected as to define the functional state. The problem will be to substitute the hitherto-unknown, characteristic time of the function, for the arbitrarily selected, characteristic time of a synapse or of a neuron (which includes all the physicochemical processes leading to impulse generation).

In the generalized coordinates system the kinetic energy $T$ is simply equal to a quadratic function of the generalized velocity. We have to remember, in this context, that in terms of impulses (following our dynamical approach), the potential energy is always zero for two reasons: i) We have seen that a potential energy can be identified only where there are no impulses and should be expressed, at least so far, only in terms of thermodynamic variables; ii) We cannot attribute a rest mass or a position to our impulses, having equated them to massless particles, never at rest in an inertial frame. Through the generalized coordinates can also be defined the generalized forces (derived from the transformation of the generalized coordinates) computed through the Lagrangian differentials of $T$ with respect to $q$ and $\dot{q}$. It is thus apparent that by also choosing $q$ to represent nervous parameters or variables other than the instantaneous rate of impulses, (provided such parameters and variables are suitably related to the system's state), one can always define the motion through the Lagrangian equations in different physiological terms, since the velocity will never be that of a displacement in physical space and time, but will remain a generalized velocity, and the momentum will not be the product of a material mass and its velocity, but a generalized momentum. For this reason, questions related to the state-representing nature of $q$ shall not be further pursued, demanding a more profound examination.

Still, the representation of the nervous function in terms of dynamic variables alone, even formalized in generalized coordinates, is not the easiest or the most practical one for defining the brain's motion. One of the principal characteristics of our nervous system, the high degree of interaction between its component subsystems, requires an approach specifically accounting for this aspect. We shall see next that the vector-based picture has more advantages, in view of the treatment of the conjoined time evolution of the subsystems.

## 8.4 States and Vector Relationships

As we have seen in Sect. 7.3.1, if we do not want to (or cannot) use energy to define the state of a system, we can resort to a different approach based on the pattern of relationships between pairs of vectors. This would again mean working with large numbers of variables (vectors) determining how their values vary with respect to each other and constructing a space of vector relationships. In the brain, even working with an extremely limited number of variables, e.g. 20, the number of functional relationships between all possible pairs would quickly become impractical, i.e. 190. Without resuming the discussion on dimensionality (Chap. 5), it is probable that for some limited purposes an acceptably precise definition of state might be based on a somewhat limited number of functions, there being a redundancy inherent in the concept of correlation between pairs of variables. Actually, there is strong experimental evidence that the behavior of different brain variables follows identical patterns in time. A reasonable approach might then be represented by the appraisal of the two extreme values of the correlation, determining which vectors are parallel (or colinear after translation) and which are perpendicular (or orthogonal); colinearity, of course, meaning the strongest correlation, either direct or inverse, orthogonality, on the other side, indicating the absence of correlation. The resulting pattern of relationships would be much simplified, becoming the pattern of distribution of only two values, 1 and 0, but, a change of the distribution is reasonably to be expected while the state is slowly developing, following the mechanism which determines the correlation. Colinearity may occur: i) Because the reciprocal interactions are such that either one of the variables can be used to define the state of the other, in a unique and unequivocal way; ii) When a third structure is driving the activity of the former two in exactly the same way. However, the efficacy of the interactions between two structures is based, on one hand, on the richness of the anatomical connections and, on the other, considering, as we have done, the structures as systems in motion, on the possibility that the forces succeed in changing the motion's momentum. It is easily appreciated – and there is some evidence to support it – that an anatomically conditioned colinearity changes less easily with the system's state, because the wealth of the reciprocal connections also tends to largely preserve a constancy of interactions under widely varying conditions. On the other hand, if the efficacy of the interactions is mainly based on the dynamics of the forces exerted by one structure upon another, colinearity may disappear whenever the structures change their activity in accordance with the state of the system. Similar arguments apply to the case of orthogonality, occuring either because of the scarcity of the anatomical links between two structures, or because the force exerted by one cannot measurably change the momentum of the other. Obviously, a change of state cannot modify orthogonality determined by scarcity of connections, while it is expected to be effective when the orthogonality is based merely on the force-momentum ratio.

The presence of one or more subsystems inducing covariation of the activities of other subsystems, thus eventually determining colinearity between the associated vectors, is more than a hypothesis. There is actually strong convincing evidence to support it. Their operation is widespread, though by no means uniform or unidirectional, so that they can modify, if not altogether change, the tendency to constancy of interactions created by the richness of connections between the structures, but can also stabilise situations based on the relative amount of driving force and momentum of the nervous motion, (see also next section).

The cross-correlation coefficient between pairs of vectors might be considered an approximate way of detecting colinearity and orthogonality. We should remember that it measures the cosine of the angle formed by two vectors. This is a statistical measure, surely, being a covariance normalized by the standard deviations, and varies continuously in the range $-1$ to 0 to $+1$, thus also revealing colinearity and orthogonality.

A non-statistical method might be represented by the "Poisson Bracket" (PB). In classical mechanics the PB represents another way of expressing the canonical relations between the dynamic variables (such as the generalized position and velocity), and is used in quantum mechanics to determine the quantum conditions, i.e. the commutation relations between variables. However, being a mere algebraic procedure, the PB is not necessarily bound to a particular situation. It is based on partial differential equations of the variables and their functions, the result being either 1 or 0, and thus is a measure akin to a correlation, again such as colinearity and orthogonality.

It is clear at this point that it is feasible, and probably useful for limited purposes, to seek a simplification of the representation of variables for defining the instantaneous state as "frozen" within a characteristic time interval. This viewpoint, however, becomes untenable when considering the temporal evolution of the system, when the changes of (theoretically) all the inter-variable relationships ought be monitored. Each of the latter may change following its own destiny, differentiated according to the particular state motion, as enough experimental evidence suggests. Therefore, every method employed for defining the relationships between the variables ought to account, also, for their continuous transformation in time (see next section).

### 8.4.1 Transformations in the Mechanics of States

The definition of state in classical mechanics is a definition of a state of motion. On this basis, we have previously discussed the types of activity in the nervous system. Now, movement in our three- or four-dimensional space implies that a body is changing its position. It is not explicitly concerned with the amount of displacement, but always implies that at each moment is enacted a change, which brings the body into different relationships with the space coordinates. This is exactly what we have been assuming for brain motion, the result of each of the innumerable subsystems unceasingly changing its own "position" (the latter term can only refer to the values of the dynamic variables, the generalized position coordinates "$q$"). In physical space, each moving material point has different spatial coordinates at an immediately succeeding time, and thus differs from itself at the preceding instant. A change of $q$ values in the brain space means the same as a change of coordinates in physical space, and since the changes of the dynamic variables consistently correspond to modifications of the physiological state, the latter can be expressed in terms of changing coordinates. Our task for the remainder of this chapter will be to define the possible ways of describing changes of coordinates or, to use physical terminology, the transformations of the reference system.

We have seen that, through the generalized coordinates, changes of nervous activity can be described in terms of the laws of motion. The Hamiltonian equations in canonical form are generally used as bases for transformations, taking advantage of the fact that the relationships between the variables, once put into canonical form, are invariant despite a change of coordinates, provided it is a linear one. This property, fundamental to every physical law, leads to useful conclusions in the nervous system too. In every physiological state, the subsystems perform many functions, each function requiring the participation of a large number of different classes of neurons. (Remember that the concept of subsystems and classes does not necessarily imply a constancy of composition in different states.) Many of the functions performed are required to remain unvaried (even if not completely) in almost every state, for instance, the vast majority of the autonomic control functions and many of the somatic functions. On the other hand, other functions seem to be more specific to particular states, i. e. they are not kept constant (are state-dependent). Thus, if the relationships between the dynamic variables involved in the functions are expressed in canonical form, and if the state changes are represented by linear transformations of coordinates, we can be sure that the functions will remain invariant after one state has changed into another. Conversely, we might assume that for state-dependent functions the transformations are non-linear.

The concept of transformation is not restricted to classical mechanics (the Galileo and the Lorentz transformations) but has found a most fruitful application, again in quantum mechanics. There, the "representation" of a vector is a way of defining it numerically, as if in a system of coordinates. At the same time, the representatives of the vectors corresponding to a state (the numbers which define it) can be used to calculate the probability of their state (it is the square of the modulus of the representative). Since, in quantum mechanics, the states can be considered to result from the linear superposition of other states, the representatives of two states can also be expressed linearly, one in terms of the other, in transformation equations, and the transformation coefficients, or "transformation functions", allow change from one to the other representation, i. e. from one system of coordinates to the other. It is clear that the same concepts apply to the nervous system, when we represent its variables as vectors, provided we assume the linear superposition principle to be valid at some level of organization (as we have done in the preceding chapter), visualising its states as real frames of coordinates, in which the vectors are in definite spatial relationships to each other, and visualising the transition from one state to another as a real transformation of coordinates.

By viewing state transformations in their minute, discrete steps, or as a continuum, as changes of a coordinate system, implies that all the variables involved in the system are simultaneously subjected to modifications of their values, such as those of degree, velocity and direction. Many of them are modified to a different value, but many others are modified proportionally, i. e. there is a true "covariation". In the brain the interaction between subsystems, assured by the enormous wealth of their anatomical and chemical links, and by their organizational principles, is a source of covariation, i. e. of the constancy of relationships. For these reasons, the evolution of the system is the evolution of the degree of covariation, which, in the brain is principally the result of "cause and effect" relationships, only subject to statistical fluctuations to a minor extent. Of course, this property does not prevent us from availing ourselves of the possibilities offered by the statistical treatment of the situation for descriptive purposes, or because the solutions of the equations are practically unapproachable. The representatives being the bases upon which are calculated the probability amplitudes for the variables of the state, the transformation functions from one representation to another can be used for computing the probabilities bound to that particular transformation. With this, we could reach a triple goal in the nervous system: i) of defining a neural state through numerical parameters (the representatives of the vectors defining the state), which by itself would be a major achievement; ii) of defining the transition from one state to another through transformation functions, another important result; and iii) of simultaneously obtaining the probability amplitudes (or the relative probabilities) of the variables corresponding to each state.

### 8.4.2 Operators and Brain States

The idea of a state transformation as the outcome of some "operation" is actually one of the fundamental principles of quantum mechanics. It is embodied in the concept of "operator", expressed with the symbols identifying operations performed on vectors (the state, ket vectors) or on other operators, with the usual constraints as to commutability. The operators in quantum mechanics correspond to the dynamic variables, or to their functions, by which the operators are endowed with immediate physical significance, apart from the obvious mathematical one (the fundamental axioms of quantum mechanics are based on the relationships between states, dynamic variables, observables and operators). Having assumed that the states of our brain can be represented by vectors, the transformation of a given vector into another is performed by a suitable operator. However, within this unique general scheme, when we deal with the temporal development of a system in "undisturbed" motion (what statistical mechanics calls "natural" motion, by definition not subjected to forces) different solutions are possible. For a continuous evolution, the general scheme of the state changes, i. e. "transformation = operator × state ket", is subject to the condition that either one must be time-dependent. The use of the Hamiltonian function $H$ as an operator produces the Schrödinger equation of motion when it operates on a time-dependent state. The operator $H$ (being equal to the total energy of the system) is a constant, thus the Schrödinger picture considers the state evolution to result from a fixed (not time-specified) operator applied to the moving (time-dependent) state ket. If we indicate the state ket by the Schrödinger wave function operated upon by the same operator $H$, we obtain for the brain what corresponds to a Schrödinger wave equation, from which the probabilities of the values can be calculated in the usual way (squares of the modulus). The Schrödinger picture is appealing, inasmuch as it connects the problem of the development of states to the probability of finding the system in a particular state (as defined by particular values of the wave function). However, one cannot say, beforehand, whether it represents the best choice, bacause we still lack any experience of computing both $H$ and state kets (or wave functions) for brain motion.

Instead of the Schrödinger approach we can use the somewhat different one we owe to Heisenberg. Considering the same system, our brain again, we can think of an operator moving in time (time-dependent), applied to a fixed (not moving) state ket. The resulting motion could be described, for instance, as a sequence of unitary transformations produced by a unitary, time-dependent operator, analogous to the displacement transformations produced by a displacement operator (or, in classical mechanics, analogous to "contact" transformations). The latter picture would seem to correspond better to what we know of the nervous system, since its physiological states are determined by a specific operator (or operators), represented by what the neurophysiologists call the "diffuse projecting systems". These operators are common to most brain subsystems, though not operating quantitatively with the same strength everywhere, nor, probably, simultaneously. Their exclusive time-dependence is not unequivocally demonstrated, on account of their interactions with the dynamic variables connected to the state of each subsystem, but there seems to be good evidence that they are the source of the time changes. In this case, a moving state function of the diffuse projecting subsystem is assumed to be the time-dependent operator on the fixed states of the other subsystems. The Heisenberg picture would not be an exact mirror image of the Schrödinger picture.

Through the Schrödinger and the Heisenberg pictures we arrive, again, at the conclusion that state and state-transition mechanisms in our brain, although conceptually separable categories, are not actually separable, and form a functional unity. Both pictures seem to be equally possible at this stage in our knowledge, and only experience may indicate which of the two is more suited for representing the brain mechanisms of state development in different situations. Thus, the fact that we seem here to prefer the Heisenberg approach does not mean that only through the latter can we expect the best solutions.

For our purposes, one of the major conclusions to which the foregoing discussion leads is that the continuous time transformations of the states (in terms of the representation of the ket vectors and of their operators) are nothing more than continuous transformations of the reference coordinates of the system. That is, they produce a moving reference system (or a moving system of coordinates), a statement bringing us a step nearer to what I consider a first sketchy outline of the theory of functioning of the central nervous system.

However, before proceeding deeper into the problems of the spaces and of their moving coordinates (postponed to the next chapter) we should try to connect the picture given by statistical mechanics in the preceding chapters to that given here by classical and quantum mechanics. Statistical mechanics conceives the states of a system as if the system moved on surfaces of constant energy, and the transitions between states as the energy increments (or decrements) necessary to jump from one surface to another. The problem of transition in quantum mechanics may be approached from the point of view of perturbations of the system's motion, produced by small fluctuations of energy, however originated. Two outcomes, then, are possible. In the first case, the perturbation is too small to overcome the difference of energy content of two adjacent states (between two surfaces of constant energy), thus the system does maintain its stationary motion and the perturbation is manifested only as a damped oscillation of the average energy level of the system. In the second case, the perturbation may succeed in forcing the system to jump between very close adjacent energy states, thus the motion at that time is no longer stationary. In the latter case, the state transition can again be conceived as being performed through a series of intermediate steps, each with a somewhat different energy content, but we cannot say in advance whether such an extremely fine resolution of intermediate states will be possible at all, (or, at any rate, useful). The concept of perturbations that we may call respectively "ineffective" and "effective" has given its name to the perturbation theory; the corresponding procedure is applied when, as frequently happens, the solution of the quantum equations of motion appears impossible. The perturbations can be appreciated statistically through the matrices of the Heisenberg quantum representation which give: i) the probabilities of the oscillating values of energy in the case of ineffective transitions; ii) the probabilities of the effective transitions between the initial and the final states; as well as, iii) the probabilities of the transitions between all the intermediate states that can be identified by seeking higher degrees of definition.

Thus, the perturbation principle also adds its contribution to enlighten us about some mechanisms of operation of the diffuse projecting systems, which, by not acting with the same intensity everywhere, induce perturbations which do not affect some of the variables (therefore, not changing their state, or their stationary motion, as some experimental results indicate), while they may simultaneously induce true changes of state in other variables. At every moment the myriads of brain variables are split into two large categories, those which are simply perturbed and still maintain a uniform, stationary motion (this category may include the ongoing, quasi-uniform motion of the subsystems); and those which, changing the state of motion, have also changed state in the physiological sense. With this, it is clear that there can be simultaneous, differentiated transformations of the coordinates of the variable space, which suggests the presence of different reference systems in the same brain. However, before we concern ourselves with such problems, we need to analyze better the real significance we attribute to the "space" spanned by our vectors, and to the changes induced therein by the state transitions.

# 9. Reference Systems for Brain Function

## 9.1 Problems of the Brain Space

In the last three chapters we have just hinted at the possibility of representing nerve function by sets of generalized coordinates, either following the laws of statistical mechanics (surfaces of constant energy, Gibbs ensembles), or of classical mechanics (canonical equations), or with systems of vectors as in the theory of quantum mechanics (state vectors, operators, the Schrödinger and the Heisenberg pictures). These methods would permit us to determine the future behavior of a neuronal population, provided its present state is known and the parameters introduced into the equations are reliable representatives of the state of the nervous activity. Such an abstract mathematical approach may not appeal much to the neurophysiologist, who is generally used to visualising in some way the position of his subsystem in the physical structure of the brain. Unfortunately, since there are a great number of dimensions, we have to abandon the idea of an actual pictorial or graphical representation. A functional description of the brain is possible only through an extended array of parameters, which may be thought of as constituting the coordinate axes of a space, the state of the system being a function of all the variables involved (the values of all the variables at each instant). Therefore, the position of the "image point" representing the state of the system will be, at some point within the space, continuously moving as time advances, following the minute changes of the variables' values. Since motion in the nervous system is not performed through displacements as in the three- or four-dimensional physical world, we would think, as neuroscientists, that the properties with which the brain space ought to be endowed need closer scrutiny. However, we shall find out that the space representing the nervous system, except for its greater number of dimensions, has characteristics which are unusual at first sight, yes, but are actually the same as those of our physical space.

Now, we assume that within that space are represented all the brain functions, which can be followed by looking at the path traced by the image point in its movement, or by recording all the changing values of each variable at successive time increments. The picture remains unchanged if we consider the variables (or their functions) as vectors. In the latter case, however, in place of a point we should consider the resultant (ket) vector, having direction and module defined by the sum of the component vectors. With so many variables and such a widespread connectivity, we must expect some degree of covariance (of correlation) between all of them, i. e. the coordinate axes will exhibit a slant toward each other proportional to the value of the correlation: in the limit, the vectors (the coordinates) may be parallel (colinear) or othogonal (see Sects. 7.3.1 and 8.4). The measure of correlation between two variables does not necessarily need to be normalized, but in many computations, such as the correlation coefficients, the covariance is normalized. It must be remembered, that in the nervous system only normalized measures of the variables, directly or indirectly representing the nervous activity, may be used for comparative purposes, because the different types of physical parameters, representing in one way or another, the nervous activity (electrical potentials from single cells, from populations of neurons, from muscles, etc.), and the variability introduced by the measuring instruments (electrodes, etc.), prevent a direct comparison of the "raw" data. Obviously, this is in line with the definition of states in quantum mechanics, i. e. the state of a system is not specified by the absolute values of its components (the modules of the vectors contributing to the definition of the state vector) but by the participation ratio, or coefficient, of each vector – a unity vector – in defining the direction of the resultant state vector.

These observations still do not give any indication of the suitable space representation for the brain. The factors to be taken into account are the characteristics of the motion itself, and the forces acting on the system. Classical Newtonian mechanics can deal with practically all types of motions of physical objects in three-dimensional Euclidean space, in which the rectangular, Cartesian coordinates constitute an "inertial" reference system. The inadequacy of the three-dimensional system is revealed when the velocities are near to the limit, to the velocity of light, requiring a fourth dimension, time, (hence, the four-dimensions, Minkowski and space-time system). Within this expanded frame occur all the phenomena of special (or restricted) relativity. Besides an increase in the number of coordinates, however, the Minkowski space does not differ from the usual Euclidean space, because it still represents an inertial space-time reference. Instead, an essential development in the concept of space and time forms the core of the theory of general relativity, inasmuch as the space-time system is no longer the "container" within which the matter of the Universe moves, but is itself constituted of time, matter, and the related energy, and is "curved" by the forces connected to matter. (In curved spaces, the sum of the angles of a triangle is larger or smaller than 180 degrees, and the shorter segment linking two points is not a Cartesian straight line but a geodesic). Still, in our Universe the effects of space-time curvature, although demonstrable, are extremely small and appear mainly in measurements of electromagnetic phenomena (such as the curvature of the star-light paths and the red shift of spectral lines). The essence of general relativity, however, is that it guards the invariance of the laws governing physical phenomena, also within a non-inertial frame. Therefore, the reference space-time must be endowed with more general properties, i. e. space and time coordinates must be curved, to a degree dependent upon the amount of matter-energy at each point; the coordinates become straight and rectangular whenever the curving forces vanish (or, for the equivalence principle, are reduced to zero); the curved space, in the limit, is transformed into a "flat", Euclidean space.

## 9.2 Properties of the Gaussian Coordinate System

The basic theory of a space such as that later required by general relativity was introduced by Gauss around the middle of the last century, and was later generalized by Riemann for curved spaces of any dimensions (the Riemannian spaces). Why I am raising the question of curved spaces for the brain's function cannot be completely clear at this stage. Now, I only can give an anticipatory justification, that is, the diffuse presence of forces differentially acting on a great number of variables, the substantially non-inertial characteristics of nervous activity, the possible relativistic aspects of brain function (to be elucidated later), all seem to make the Gaussian system of coordinates the best candidate for a brain space. In the Gaussian general system of coordinates, the axes can be curved to represent a curved geometry, but if the situation allows the measurement of sufficiently straight small segments, then, in these small vicinities, the Euclidean geometric relationships (the geometry of "flat" spaces) are approached again. In Gaussian space, the distance between two points is measured through the so-called *ds* metrics, upon which the generalized theorem of Pythagoras is based, introducing in its calculation the coefficients representing the degree of curvature, (or of slanting) of the axes.

Since the Gaussian system of coordinates is fundamental for the future development of our theory, we will now proceed to outline those of its properties which are most relevant to our concern. It is a system of a very general type, accommodating any number of curved or straight axes. Each axis is identified only by a number, and the position of each "cell" of the mesh is defined by the numbers identifying the axes enclosing it, while the measures of the sides of each cell are defined by the metric coefficients of the *ds* metrics. It is this extreme generality which permits the Gaussian system to adapt to every conceivable condition, such as to the number and nature of coordinates, and to the distribution and intensity of forces (which are very uneven in the nervous system) so that the "local" conditions in each cell of the Gaussian representation may widely differ from nearby ones.

Let us see what can we make from the Gaussian coordinates. As was said previously, the correlation between the behavior of any pair of variables can hardly be constant, on account of the fact that a physiological state is not a frozen, fixed condition, but a steadily evolving situation, in which the values of the variables are continuously changing, some increasing and some decreasing as the functions of that state are performed, while different aspects of the outward behavior are appearing and disappearing. However, all this means that our Gaussian representation of the vector space, being bounded by continuously slant-varying (and, perhaps, curvature-varying) coordinates, is continuously subjected to deformation. Let us consider the latter aspect. We have been talking in the last chapter about the physical laws (such as the laws of motion) which remain invariant, if in canonical form, under a number of linear transformations (changes of coordinates). Now, I am assuming that an invariance is also maintained when the nervous system transforms its physiological state (a change of coordinates being a change of state), whatever this transformation might be, as is assumed in the theory of general relativity. This can be achieved in only one way. Since within one and the same physiological state, the space bounded by the variable vectors in the Gaussian reference is continuously undergoing deformations, the consequence is that also within the same state some continuous transformations of the coordinates' values (see the preceding chapter) must be suitably operated, in order to maintain the invariance, otherwise every function related to the values within that state is likely to change. As a matter of fact, any function that the brain performs in a given time interval can be represented by a segment of the path covered by the image point in the same interval. Therefore, it is expected that the size of that segment remains unvaried if the function represented by that segment is invariant. Hence, it is the system of coordinates which must ensure that the values may retain an invariance notwithstanding the deformations of the set of axes. In other words, the representation must be such that the unit measure of the coordinates may vary together (covary) with the deformation of the space, (the latter proposition deserves a more accurate discussion in view of the subjective phenomena in different brain functions, to be dealt with later). The measure of distance between two points – actually representing two neurological activities or events – separated in time or in space, or both, remains the same only in Gaussian systems, where the unit length defining the measure depends upon the linear "dimension" of the cell, (a dimension which is itself state-dependent). Only in this way may the value become a constant of the space, and whatever physiological significance is attached to the image points and their measures also remains invariant. It is clear, conversely, that through the invariance of the values we assume the functions representable by distances between points in the Gaussian reference system to be invariants of the brain activity.

## 9.3 Invariant Functions in the Central Nervous System

The existence of invariant functions in the brain is a matter of fact and should not need to be demonstrated, when the term simply indicates the nervous control of some stable vegetative functions (based, mainly, on self-stabilizing, homeostatic, and reflex circuits). However, upon closer examination, the invariance seems also to be a property of very complex nervous processes such as perception, recognition, organization of movement, abstraction, logical thinking, organization of behavior, etc. Through our personal experience in everyday life, we know, for instance, that our recognition of a face, a voice, a contact, an object as a whole, is not essentially influenced by a change in brain state, such as the changing degrees of alertness during waking, the states of rage or quiet, of high spirits or of low mood, and so on. Moreover, there are states with very different levels of consciousness, such as, for instance, sleep, in which our dreams are populated by images that we can recognize as everyday images of faces or of objects, we perceive well-known odours, we hear sounds and words which are readily recognized and correctly attributed to known persons, and remembered afterwards. It may be assumed that this recognition, with other functions which are correctly performed in different states, constitutes an invariant among the operations of the brain, representing something fundamental that we cannot do without. We have only to think about what would become of (not only) our social life if we were to misjudge faces, objects, sensations, or situations as a whole, with every small change of the state of our brain. That the sensory experiences, the faces, the sounds, might not be aroused in our brain from without, but actually produced within, is immaterial: the ability to correctly recognize something, somehow extracted from "memory banks", seems to represent the routine. In every state or situation the recognition would be performed (after an analysis) through comparison with similar "patterns", with some kind of memory content (at least, this seems to be the prevailing interpretation), and must be founded on the invariant properties of brain functions.

The transition from one state to another brings about differentiated changes of all the coordinates (the variables) and, correspondingly, may produce changes of the correlations between them (measured by the usual statistical methods). As substantiated by recent experimental evidence, the average cross-correlation coefficients between some variables are statistically, significantly different in the extreme physiological states of sleep and arousal. The state-dependent differences in orientation of the vectors, or the different modules do not by themselves define whether the transformations involve rectilinear or curved coordinates (the coordinates may be either curved or rectilinear without changing the significance of the transition from one state to another). However, the point is that in order to assume an invariance, only in the cells of the Gaussian reference system can the unitary measure of the sides either shrink or expand as a function of the state, and the orientation of the axes be a function of the same state.

Unfortunately, a complex picture such as that of the Gaussian coordinates cannot be put on graph, except by using two coordinates at a time, and even this would become imprecise, were the coordinates curved. We must try to mentally visualise a reasonable picture of the situation. Our brain can be represented by the myriad of its "partial functions" (such as sensing, moving, etc.), those simple phenomena which together build up what we identify as an "integrated" function. We are also unwilling to consider the latter as independent, rather than as part of more complex nervous phenomena ultimately giving rise to behavior. We may consider each partial function to be defined within one cell (belonging to the whole space) bounded by a number of coordinate axes, each representing one variable involved in that function. The ensemble of these cells, all irregularly shaped, is continuously undergoing transformations as the state develops, and slowly or quickly the cells are transformed into others in an endless succession. This cerebral "mass" (not the physical brain, but its representation!) would be like Einstein's "reference mollusc" (a term he used to refer to the Gaussian system representing our physical four-dimensional space), which can take on any shape in its different regions, still ensuring invariant values within each cell, the principal and characteristic functional property of the Gaussian "mollusc".

Actually, the Gaussian system of coordinates has found in the general relativity theory the physical reason for its existence. In this theory, matter and energy automatically "adjust" to their local amounts the values of space and time, in the different parts of the Universe. By abstraction, in a completely empty space, or rather, sufficiently far away from matter and energy, where the forces are negligible, the coefficients of the *ds* metrics define a diagonal matrix, i. e. flat, Euclidean geometry is valid (since only matter and energy curve the space). The important point here is that, if we accept the Gaussian reference system for brain function, we must also necessarily assume, to assure the invariance of several of its measures, that in our brain too there is an "automatic adjustment" of the space metrics, as the Gaussian space is deformed while the forces transform it from one state to another. This constitutes the basic assumption of our theory (to be carefully tested, of course, by experience). On operational grounds, the idea I am trying to convey is that the mechanism which induces the transition from one state to another also produces, at the same time, together with the differentiated increase or decrease of the variable values (depending on the variables and on the state), the change of the metrics of the sides of the subspace bounded by those variables. This causes a deformation of the space coordinates (thus, of the cells of the Gaussian mesh) but allows, by that, the maintainance of the constancy of the quantitative relationships between given variables, i. e. a constancy of correlation. It is the same principle of covariance as in the Einstein's theory of the physical space (though applied to subspaces of more dimensions), which ensures in the brain the invariance of the measures.

In physics and statistics the term "covariance" indicates two aspects of the same concept: on one hand, the covariation of the equations expressing the physical laws when the reference system undergoes transformations; on the other, the statistical measure of the covariation of the system's variables. The essence of the concept is "maintenance of the quantitative relationships between events, variables or their functions", and therefore, of their relative measures in different states (or reference systems). Since the statistical measure of the covariation of two variables is the coefficient of correlation between the two, the latter defines the extent of their interdependence (in statistical, not in deterministic terms). Now, as repeatedly stated, the value of the correlation coefficients, the way they are measured in the nervous system, is influenced by two factors: i) the extent of the direct and partly indirect reciprocal connections between the two structures; ii) the influence of other structures connected with the former two, increasing or decreasing the effects of their reciprocal connections. In the theory of general relativity, matter energy completely defines time and space through self-interacting fields obeying non linear equations. Our present imperfect knowledge of the brain does not allow predictions about the linearity of relations. The computation of the statistical correlation coefficient assumes linear relationships between the variables. Its extreme values, $+1$ and $-1$, indicate a complete statistical interdependence. However, the latter could also be interpreted in a deterministic sense (and only the physiological situation can tell whether it is actually possible) because these extreme values could indicate the highest possibility of interaction between variables, forces and fields. This situation, however, should require non-linear equations, as in physical space, in order to completely describe the brain phenomena.

### 9.3.1 Which Functions Are Invariant?

Now, let us examine closer the problem of the invariance of the functions in our brain. We shall assume, for this purpose, that by a change of metrics of the Gaussian space mesh, the distance between two points within a cell remains invariant. As we said before, any physiological property bound to this measure could thus be assumed to remain invariant too. This should be a different situation from the usual one in experimental psychology, say, the constancy of perception (better, of judgement) when an object is presented under different angles of perspective (as produced by rotating a rigid body). We can recognize, except under some limiting conditions, the nature of a solid, either a purely geometrical form, or any other object of everyday experience. In the latter instance, it is the shape of the object that changes, at least in its gross appearance, whereas in the former case of the changed Gaussian metrics what is changed is the "condition" of the brain. However, if these two judgements by the brain were ultimately to be based on distances between a number of points (i. e. on the relationships between parts) in a Gaussian space, the mechanism of the invariance could be the same in both cases.

Still, there are two clearly separate, albeit not always precisely definable, situations. Even in the task of judging the characteristics of images, there seem to be invariant and non-invariant outcomes. We can always recognize our brother's face, our car, dog or house (but also the sound of a voice or a scent) whatever state we are in, wide awake or dreaming. It is a kind of "all-or-none" recognition, the identity of the object of sensory experience being taken as a whole, not requiring accurate analysis (everyone knows that, after some practice, few cues are actually used for such judgement). Nonetheless, this is strictly, or mainly, a "qualitative" judgement, inasmuch as it is founded on the instantaneous pattern of the characteristic qualities of space (the spatial arrangements of points, the distribution of light and shade, of colours with their hues, etc.) or on the pattern of relationships of the sound components, etc., for which a comparison pattern (stored somehow and somewhere) is assumed is available, ready to be matched with this. So, the judgement would actually be the result of a comparison of a new with an older pattern, and the invariance of these judgements is hardly to be questioned.

There are other situations, i. e. when we want to measure actual distances, dimensions or intensities, (of lights, sounds, odours or touch), when the invariance of the judgement seems to vanish. Misjudgements dependent upon attention decay, dizziness, intoxication, fatigue, sleep, and mood are an everyday experience, well known in factories, in the traffic, in all control systems where Man is a part of the control chain. These judgements must be regarded as state-dependent. A typical judgement (perhaps the most important in practice) which seems to be state-dependent is that of time length, or time interval. The state of the brain strongly influences our subjective time experience. Pleasant or unpleasant situations, dreams, excitement, boredom, mood, fatigue, and hunger typically seem to meddle with our "personal" time, so that the minutes and hours seem to be running either too fast or too slow. How this personal experience of misjudged quantities, the variability of judgement, complies with the invariance of what we have called the "matching" of qualitative properties needs explanation. Why can we recognize in almost every possible situation and brain state that two crossed rods are two crossed rods, whereas we are likely to judge their actual length incorrectly if there are no reliable external clues? A reason is not to be found in the intrinsic complexity of the images presented to brain processing for quality identification and for actual measuring.

### 9.3.2 Time and Space Measurements

There could be a still-hypothetical, but reasonably sound and simple explanation for that discrepancy, fitting the principle of measurements in Gaussian space. We are living in a four-dimensional world, in which the fourth dimension, time, is on equal standing with the dimensions of space, although clearly not behaving exactly like them. In physical space, the time, measured by our clocks, is influenced by the factors (forces) which curve the space, i. e. matter and energy, and therefore time and space are correlated (are "covarying"). However, in the reference system of the brain the time coordinate represented by the clock time (or physical time) cannot be influenced by brain forces, and thus would not be correlated with the other coordinate axes, while the slant (or the curvature) of all the axes will change to a degree dependent upon the state. Clock time will thus be in a continuously changing and state-dependent relationship with the rest of the brain space coordinates, because since it cannot be influenced by the brain state, it does not "adjust" its value (following relativistic logic), as the other variables do at every change of state. This would seem to draw a line between situations which do, or do not, involve time. Whenever time is not involved we could expect our judgement to be invariant in the different states. The typical case mentioned is the recognition of faces, objects, sounds, odours, and the like. All these operations take into account the simultaneous and instantaneous pattern of relationships between all the components of the stimulus, somehow matching them with similar patterns that are already present and available for comparison. Where time is concerned, we seem unable to set up a comparable situation based on simultaneous comparisons. In other situations too, in which we have to make measurements, i. e. really quantify extensions and intensities, we can easily assume that there cannot be ready-made matching measurements in our brain, so the comparisons have to be performed with measuring devices from the outside world, not from within the brain. Our assumption, thus, can be that the judgements which appear to be state-dependent are so because real physical measurements of space and time are involved, representing "absolute" interval measures (of whatever type), of which there seems to be no way of keeping a record (memory) in our brain. Therefore, time interval measurement is a typical state-dependent function of our brain, as much so as every other interval measure of space, of intensities, etc.

Thus, we reach the necessary conclusion that the invariant functions should be based exclusively on the relationships of the elements of space (or of other categories) among themselves. This, however, renders all the more puzzling our inability to make invariant measures in space. After all, we could assume that in the case of real interval measures we are again confronted with the task of establishing relationships among space elements, such as the distance between points (which forms the basis of "matching" recognition) with reference to some "standard space measure. However, we have no hints of the existence of such an absolute space reference. It might be assumed that the variability of the measures is to be attributed to time, which contributes the fourth coordinate in every measure within the Gaussian space. With regard to this hypothesis, simple calculations with the Lorentz transformation formulae show that the influence of the time non-covariation would become detectable only when relativistic velocities and distances are involved, which is never the case in our everyday life. (However, see Chap. 12 for the relativistic aspects of brain function.) Therefore, the only acceptable hypothesis is the one just hinted at, i. e. that the invariance of a function (such as the correct recognizing of a face) is based on "whole" matching of the internal relations, no matter how large or small the components of the pattern (say, a large or a small picture of the same subject). In order to obtain the same invariance for absolute measurements of interval, it should be possible to perform matching with "on scale" measurements of the same intervals stored somewhere. Moreover, in a complex pattern there are very many (sometimes innumerable) interval measures, and each is congruent with every other, (as a matter of fact, each one supports the others), whereas for a single interval, line segment or time, there are no other congruent measures which can offer a clue. Actually, the presence of false clues is the basis of the innumerable optical misjudgements (illusions) described in every book of physiology or psychology. In the absence of such collateral clues, an invariant interval measurement could only (if ever) occur in the case of a measurement made in exactly the same state and situation in which an exactly equal reference measurement was stored. In all the other states, adaptation of the brain Gaussian space metrics to the state would not produce a measure proportionate to that definite interval of the outside world.

Let us now turn our attention more specifically to the parameter "time". In accordance with the non-congruent behavior of time with respect to the other parameters of space, there is experimental evidence that time length discrimination is influenced by the brain state, probably in a systematic way. The time between two reference stimuli has been reported by test subjects to be longer when in that interval they were not active or were performing boring, dull tasks, whereas it was judged shorter when they were kept busy in interesting work requiring mental concentration, and while receiving and processing complex stimuli. This means that in the former case something we might call "personal" time was running faster than the clock time, whereas in the second instance personal time was advancing more slowly (always by comparison with clock time).

The terms "run" and "advance" applied to time unduly emphasize an independence of time from other physical variables, but the two examples quoted hint at a possible profound interdependence of personal time and the other variables. In the theory of relativity, clock time is slowed down by acceleration and by matter energy (the source of gravitation). The same thing seems to happen with personal time, since the greater the involvement of neurons in the responses to stimuli (interaction with the outside world) and in complex internal mental processes (stimuli from within the brain), the more neural activity can be likened to accelerated motion, that which slows clock time in physical space. What has changed pace in the brain, and is related to time, is not its metabolic energy, but something more strictly bound to the brain's dynamic variables, as shown by the fact that strongly diversified situations of general body energy metabolism seem to have no influence on time judgements. It cannot be determined whether all the brain subsystems are equally capable of influencing personal time. It may be possible, for instance, that the two types of judgements, simultaneous comparisons and interval measurements of extent intensity are performed in two distinct regions of the brain, say the two hemispheres, the right one for the first type, the left one for the second (as accumulating evidence would suggest). However, this would not change the problem with respect to the state-invariance and state-dependence of the measures in Gaussian space, if the time measurements are made in the same side in which the extent intensity measurements are performed.

Now, however, the situation as far as the time coordinates in the brain Gaussian reference system is concerned, appears profoundly different. If, in place of clock time, we assume that the time axis in the brain space may be represented by the personal time, then, during state changes, the same "adaptation" will occur for time as well as for the other coordinates induced by the same brain forces. Thus, we may realize in our brain space the same situation as in physical space, where every measure of space and time is invariant, provided we assume that for every subsystem and its states, besides a "local" geometry, as expressed by the *ds* metrics, there is also a "local" time. Personal time would vary with the brain state, which clock time cannot do. (It remains to be demonstrated that there is an exact correspondence of personal times with all possible different states.) As a consequence, the personal time would represent a constant for each state, a true state variable. In turn, this also entails that the measurements performed on all the coordinates (time included) of one and the same space should give different results when judged by an external observer (one at rest in a different reference frame). But with these problems we have already encroached upon the field of relativity, and we shall discuss a few aspects of that theory later on (Chap. 12).

I should not like to push further in this chapter the assumption that the personal time is made to vary together with the other space coordinates during state transitions. The data in support of the assumption, though suggestive, are still too few and ought to be completed through a methodical search for correlation between brain motion and change of personal time. However, the problem of the state-dependence of time interval measurements must at least be viewed from another (perhaps more realistic) perspective. We can safely assume that time judgement is made only when the forces which have induced the change of state have come to an end, or have become stationary. (We said that these forces induce a series of short-lasting accelerations, attaining a stationary level only after a transient period during which their effects accumulate.) This assumption is quite reasonable, since every measure requires stationarity of the measured object, and we know of no reports of awareness of an "accelerated" time. (Perhaps it has never been subjected to an experimental test, although it would be of extreme interest to ascertain whether it is possible at all.) The only accelerations we seem to be aware of are produced by the forces acting on our spatial coordinates, but for them we have receptors in the body, whereas we seem to have no receptors for time. (However, a "feeling" of slowed time, something very much like a "sensation" of it, has been reported by Eccles, and apparently by others, in emergencies, when there must have been a strong emotional involvement as well.) The time interval judgement could be dependent, thus, upon a steady state "measurement", for instance, of an energy parameter (the metabolic one, or the Hamiltonian) or other dynamic variables, but, being performed when the forces have reached a stationary level, would not be directly influenced by the gradients of energy which have operated the change of the other variables. In this connection, it might be interesting to recall that time and energy are "conjugated" variables in quantum mechanics, since their PB is 1; see Sect. 8.4.

We must concede that many of the difficulties met while investigating the matter of time judgement in different states are due to our poor knowledge of the states of the brain, besides the few clearly discriminated through behavioral parameters, which refer, after all, almost to only two levels of consciousness, low and high. Actually, what is needed is a large choice of tests for a wide range of brain states, discriminated not simply on the basis of EEG spectra or behavioral clues, such as reaction times, but principally on the basis of their dynamic characteristics, that is, of the dynamic variables of brain activity. These tests should create situations corresponding to those of relativity, in which time dilation and space contraction occur as result of periods of acceleration. Again, these are problems which would lead us more deeply into the relativity theory to be more closely considered later on.

## 9.4 Closing Comments

The explanations advanced in the preceding sections may be suitable for the invariance of perceptions, of matching judgements, etc. but not, however, when more complex functions are involved, such as abstraction, logical thinking, ethical judgement, organization of behavior, and others. The fact that in the previous examples we were dealing with spatial and temporal judgements should not make us forget that the operation is performed in a brain space, representable through a Gaussian reference system, and that each of the myriad of functions is represented within one of the Gaussian multidimensional cells. In principle, the same mechanism of invariance ought to be expected for "every" function, if we could assume that an automatic adjustment of the values of "every" parameter accompanies "every" state transformation (as everywhere in our physical Universe there is adjustment of the local metrics). That, in the practical situations of life, our acts do not seem to be ruled by this type of inescapable logic is not necessarily due to a lack of basic invariance of the fundamental brain processes. A change of an expected behavior, e. g. involving ethical matters, does not imply a change (that is, the non-invariance) of the ethical judgement. This is a fact which every individual has experienced. The decision to behave consistently with the ethical judgement is not uniquely determined by the judgement itself, and therefore is independent of it inasmuch as it is determined by a host of different, and often oppositely directed drives. (This matter will be further pursued in the Appendix.)

Reducing the fundament of all the invariant cerebral functions to a matching of patterns of relations between the elements involved in the functions is an attractive idea, but it remains to be demonstrated whether it is appropriate. However, I cannot at this moment see immediate reasons for rejecting the hypothesis. In the general case of invariant functions, not simply those involving comparison of physical images and the like, a pattern comparison need not be restricted to a single "picture", but may well require time sequences and time dependence of the patterns, provided absolute measures of time intervals are not involved. On this basis we may conceptualize, for instance, the invariance of motor functions consisting of sequences of physical acts, and therefore, more generally, of nervous commands which may be performed in the correct chronological order, though with a slowed down or an accelerated "tempo" in different situations. A continuous comparison of patterns might take place during the development of a complex function such as rational thinking, where at every moment the need arises for a decision (a choice) between branching paths of the thought. We may conceptualize thought itself as "constructed" through the congruence of single elements progressively building more complex patterns, and the judgement of congruence as based, essentially, on the fitting together of patterns. However, there is no reason to prolong now the discussion of a matter still awaiting approval of its fundaments.

We have begun this chapter aiming only to find a system of coordinates, i. e. a system of representation of the brain functions, and we have ended up with an analysis of its properties which has brought us to produce something like the kernel of a real theory of functioning, however incomplete and vague it may still appear. This theory, though considering a fundamental characteristic of the brain, the constancy of its functions, remains incomplete since it has to be carefully tested for other properties, although it certainly satisfies the main one outlined at the beginning, the capacity for acting as a whole. The latter property, which is based on the widespread maze of reciprocal connections, and is functionally made evident as diffuse cross-correlations between the variables, is visualised in the Gaussian reference system, where is expressed through the principle of "covariation" of the variables, induced by the mechanisms which control the brain state.

Since the framework for neurophysiological phenomena provided by the theory just sketched is a general one, we shall now aim to gain a preliminary indication of the nature of the brain operations leading to "covariation".

# 10. The Continuum in the Central Nervous System

## 10.1 A Few Basic Ideas

In order to proceed, it is better to briefly review where we have arrived so far, and this will clarify our subsequent steps. A Gaussian reference system better represents the large number of variables with their extended multitude of cross-connections, which build a space that is probably not Euclidean. Moreover, a change of the brain's physiological state can be assumed to correspond to a transformation of coordinates, resulting from the changed variable values, the Gaussian representation permitting any change of reference system. Finally, a change of space metrics induced by the forces which cause the state transition, (as proposed in Sects. 9.2 and 9.3), is the basis for maintaining constancy of relationships among groups of variables, thus assuring in the Gaussian representation the invariance of the functions associated with those variables. So, the usefulness of the Gaussian reference system is that it represents the frame of reference in which invariant measures in the variables' space become possible when the transformation from one state to another changes those variables involved in the function that should remain constant. As we have seen, what is called the state of the brain is actually not a stationary situation, but one in continuous motion, or evolution, through an endless sequence of elementary steps. Different representations of state evolution are possible, including unitary transformations in quantum mechanics, contact transformations in classical mechanics, perturbations, moving (time-dependent) states with fixed operators, and fixed states with moving operators. There are purely statistical solutions, such as the computation of the probability of states and transitions, and deterministic solutions (such as that of the Lagrangian equations or of the Hamiltonian canonical equations). What is needed now is the unification within a single frame of all the mechanisms of state change, yet to be represented in Gaussian space.

An immediately appealing image of the Gaussian system of coordinates is the "reference mollusc" (given by Einstein for physical space), a volume of space time which can shrink and swell, and can develop local bulges and pits, as functions of the local forces acting on the space and deforming the sides of each reference cell. The reference mollusc emphasizes the continuum in the representation of the central nervous system. Within the continuum the effects of forces are seen as changes of the local metrics, which result in deformations of the coordinate system, with or without change of the correlations. We may go further, and borrow the concept of fields of forces, since the latter are throughout the continuum.

So far we have not needed to bother much about the nature of the forces, but now we shall consider that they can be represented as operators (this actually restricts the choice to those which can be represented as dynamic variables, as in quantum physics) acting upon the states of the different subsystems. The concept of a field of force in the nervous system is supported by the widespread connectivity of the neurons in progressively larger circuits, through which forces can be exerted at large distances from a source. Given that the sources are innumerable and interconnected, their respective fields are interlaced, and are not necessarily homogeneous or isotropic. Since Gaussian space is deformed by the forces present, one obviously concludes that in the absence of neural forces the Gaussian coordinates would be orthogonally fixed, and that the space could only have Euclidean properties. It is easy to perceive how in the brain we are led to pursue, step by step, some of the essentials of general relativity, (of which more will be said in the last chapter).

In the treatment of the situation to be considered next, we shall have to take heed of the existence of two strictly connected fields: the field of the cross-relations, which is expressed by the strains, or deformations, of the Gaussian space, and, (since the actual, functional relations are the result of a force action) the field of the deforming forces. However, it is always more expedient, instead of directly measuring the forces, to measure their effects at each point in the nervous system, although the two factors, strains (deformations) and stresses (forces) could be lumped together under one formulation (see also Sect. 10.3).

## 10.2 Factors Shaping the Fields of Force in the Nervous System

In the nervous system there is a wide variety of causes of state change, accounted for by forces: the quasi-natural evolution of the system (see Chaps. 6, 7 and 8), sensory stimuli, changing internal conditions of the body, pharmacological agents, fatigue, stress situations, etc. There seem to be other causes, however, which at first sight could not be considered on the same footing – except by making unjustified assumptions or unduly stretching the concept of force – e. g. all those changes of activity, (thus, changes of state), consequent upon alterations of the anatomical structure. However, by considering the matter a bit more closely, we find that the latter situations, though apparently so different, can actually be treated on a common basis with all the other ones.

As was previously said, the fields of force exist because there are connections, both direct and indirect, between the structures in such a huge amount that every structure can be considered to maintain contacts with, i. e. to exert forces upon, large areas of the brain. Obviously, the intensity of the forces will be different in different directions (so, the field will be anisotropic), and we can conceive of lines of force within each field. Although this may form a complex picture, it is easily understood that even what we have called the elementary steps of change (elementary states) will be characterized by different distributions of the lines of force, or at least of the intensities of the fields in different regions. It is all the more obvious, then, that alterations in the anatomical substrate (the neurons and their connections) will cause a different distribution of the fields, of their intensity and of the related lines of force. Changes in the anatomical substrate are actual damages to tissue (from ageing, circulation defects, degenerative and infectious diseases, trauma, etc.) so that knowledge of the fields' deformations and of their geometry should indicate the site, the size and the functional extent of the damage. This brings us back to Chap. 4, where we briefly treated the approach of localized lesions in the brain, and the analogy with a rubber sheet was introduced. In that instance, the tension of the rubber sheet and the related lines of force acquired different values from point to point, and a different distribution, as a direct consequence of the sheet's damage. What I am proposing now is no longer an analogy, but the real functional consequence of a structural alteration. In nervous tissue, in which the connections convey excitatory as well as inhibitory influences, the selective destruction of any structure, even if not complete, necessarily brings about an imbalance of the excitatory and inhibitory effects on other structures, very often revealed by gross changes in physiological manifestations: increase or decrease of muscular tonus, synchronization or desynchronization of the brain waves, tameness or aggressiveness, etc. It is clear that these functional alterations can be induced only if the lesions have produced in the fields of force a different distribution of values in different directions, which could be experimentally revealed, for instance, by the changed values of the correlation between pairs of the remaining, intact structures.

Within the same framework we can also represent the effects of stimulation (both natural and artificial), since the stimuli either produce anew (or increase) excitatory effects, or enhance (or suppress) inhibitory ones. Coherently, the effects of drugs can be visualized in the continuum through the changes induced in the distribution of the forces and of their fields. It is actually in this continuum that we can better understand how the nervous system may work as a whole, once we regard the neural connections, building up the anatomical substrate of its organization, as constituting many preferential paths through which the forces are, so to speak, radiating from many centers at a time, and exert their actions on nearby and distant structures. By visualising nervous activity as a continuum, we unify and make understandable, through a common representation, the diversified phenomena occurring in the nervous system. Since this procedure offers many opportunities for its exploitation, it is worthwhile to explore how the theory of continua has worked its way forward in the physical domain, and which of its methods and principles are suitable for a theory of the central nervous system.

## 10.3 A Fundamental Concept: The Tensor

Once we have accepted the idea of a continuum for nervous function, we have to apply to it the basic approach used everywhere in physics to account for the transformations which the elements of the continuum are undergoing. The approach uses a concept, simple in itself, though comprehensive and, in principle, without constraints as to the number of variables involved. This makes it a theoretically ideal tool for coping with the multitude of simultaneous events in the brain, although, in practice, the dimensionality of the system may recur to make the procedures cumbersome. There is a reasonable expectation, however, that with automatic computation the task will not be altogether impracticable.

Let us see, then, what basic concept we must assume to be applicable to the nervous function. The problem we are faced with is not that of motion in a straight line, or of rotations of rigid bodies in space, but that of progressive deformations of the continuum. In all the latter cases, however, similar basic rules for the transformations are needed. Continuum mechanics has condensed these rules with the help of an invaluable tool, the mathematics of tensors, developed from vector mathematics. The neurophysiologist reader is warmly recommended to become acquainted with the underlying basic concepts by consulting some simple monograph (or revising his undergraduate study of it), because only a very sketchy outline of the principle can be given here.

A tensor is an operator (the term tensor is borrowed from the study of elasticity, indicating that operator which defines the tensile stress per unit area of a body), i. e. the agent of an operation performed on vectors to transform them into other vectors in a linear way (linearity is a prerequisite for tensor operation). In the tensor concept are defined the rules by which the operation is performed. Indicating in itself only some linear operation, the term tensor is applicable quantities of different order: a scalar is a zeroth order tensor, a vector is a first order tensor, while the second order tensors form a large class including the most useful tensors in the physics of continua. There are also higher order tensors, appearing less frequently in the three-dimensional world, but probably needed in the multidimensional space of the brain.

In order to better understand the way the tensors operate and are constructed, let us confine ourselves to a classical example in the three dimensions of our world. Imagine a solid, e. g. a cube, to which a force is applied in a direction perpendicular to one of the planes of the space. In addition to the stress normal to that plane, stresses are generated tangential to the other two planes, and likewise for the stresses normal to the other planes. This gives us a total of nine quantities (or "components") describing the stress situation in that volume of space, expressed as partial differentials of the stresses from each plane to the others. The so-called (second order) stress tensor may represent in a shortened notation (for instance, a capital letter with higher or lower indices) the whole set of nine differentials. The latter, however, can also be explicitly expressed as a $3 \times 3$ matrix, either as differentials or substituted by symbols and indices for the direction in space to which the differentials are referred. The tensors can be operated upon (added, subtracted, multiplied, differentiated, contracted and expanded) to give rise to other tensors of the same, or of lower or higher order. However, for our purposes, we should never forget that they are essentially linear operators which transform vectors into vectors.

The paramount value of tensors, for physics, is that any system of relationships that can be expressed in tensorial form is maintained unaltered in other reference systems. We have already discussed the invariance of the nervous functions in relation to the Gaussian system of coordinates. It ensues that in the Gaussian system the nervous functions can be expressed by tensor operations, from straightline motions to rotations, from contraction to expansion, from rectilinear to curved axes. Therefore, it is through knowledge of the tensors that we may gain more insight into the general mechanisms of brain function. Of course, it is not expected that they will produce the answers about the primary, innermost sources of brain's functioning (such as, for instance, the innermost origin of our thinking capacities), but they will bring us nearer to them, by helping us to understand "how" nervous events are produced and processed, by offering the means of handling complex situations symbolically, of working with them as mathematical functions, therefore of grasping the "rationale" of the complex of relationships between the innumerable parts of the brain. We have here not a tool "a posteriori", like statistics is, but, I believe, the representative of the primary, fundamental "logic" of the system. This seems the most we can request without trespassing across the border of metaphysics.

### 10.3.1 The Metric Tensor

Discussing Gaussian space, we said that transformation of coordinates with covariation of all the coordinates' measures, including time, assures the invariance of the functions represented in that space. This simultaneous variation is inherent in the tensorial character of the operator of transformation. What we have called the automatic "adaptation" of the measures of the cells in the space is simply the result of an operation performed by the so-called "metric" tensor. This tensor, operating the scalar product of normalized vectors in Euclidean space, is a logical application of the concept of general tensors, those operating transformations of coordinates. The metric tensor is also "fundamental", a term emphasizing its importance (particularly in the theory of general relativity). It measures the sides of every cell of our physical, four-dimensional space (remember that the latter is a Riemannian space, a generalization of the Gaussian one). In that space the curvature is defined by a differentiation of the metric tensor, the Riemann-Christoffel tensor, which is contracted in the Ricci tensor, and by a curvature scalar (resulting from contraction of the Ricci tensor). Since in Einstein's theory the curvature of the space is related to the amount of matter and energy present, the degree of curvature is a measure of the force transforming a "flat" Euclidean space, in which it is possible to find rectilinear and orthogonal coordinates, into a Riemannian space in which no rectilinear reference is possible, a curved space.

In the four-dimensional Minkowski space we live in, the components of the metric tensor number 10 in all, (not 16 as could be expected), since, this tensor being symmetrical, the components of the matrix are pairwise equal on the two sides of the diagonal. The components are reduced to only the 4 diagonal ones when the space is free from matter (rather, when the system is far enough away from matter for the gravitating effects of the latter to become absolutely negligible). The consequence of the disappearance of the forces which curve the space is that the latter becomes "flat" (Euclidean). In this way, the off-diagonal components of the metric tensor can be considered as the "spies" of the presence of forces, since they are a measure of the effects of the forces acting on the coordinates of the space. As a matter of fact, if (because of the absence of forces, and a suitable transformation of coordinates) the off-diagonal components are null, or their value becomes constant, then the Ricci tensor vanishes (and with it the curvature). In relativity theory this corresponds to the transformation of an "accelerated" system into an "inertial" one. If such a transformation can be found, then the forces are "fictitious", otherwise they are "permanent" (see, however, Chap. 12 for these relativistic aspects).

### 10.3.2 The Metric Tensor in the Central Nervous System

It would seem that the properties of the metric tensor are just those needed to describe the situation in the nervous system. We had to resort to a Gaussian representation in order to meet the necessity for "covariation" in the presence of a continuous modification of the shape and, probably, of the volume of the space. It is useful, to avoid misconceptions, to reiterate that I am using the term "covariation" simply for indicating simultaneous variation of the variables, (thus not implying the concept inherent in the term "covariant" applied to tensors). Also, what we have called personal time would participate in the overall covariation of the brain variables, since it seems to be differentially affected by different types of brain activity. Therefore, the Gaussian space time of our nervous system could be either a curved one, whenever there are forces acting within the brain structures (an event occurring regularly since there is no moment in which the brain is at rest and not involved in complex processes), or a flat one with Euclidean properties if these forces are null or in a steady state (i. e. have become constant, the only event likely to occur in the place of an absence of forces).

The correct way of constructing the cerebral metric tensor in each state requires finding the characteristic time for each function of the state, then, averaging the variables' measures according to the respective time, and, finally, building the components of the metric tensor with the partial differentials between all the variable pairs – including personal time as the time coordinate of the system. Probably, at the beginning the procedure will be based mainly on trial and error for the determination of the characteristic times and of the personal time. However, a great deal of patience has already been required for our task, and every successful step with one of the variables means a better insight into the general mechanism of the functions. The procedure, though perhaps painstakingly long, will never be a dull one.

On the other hand, statistical procedures can help in this matter, at least at the beginning, again through the cross-correlation coefficients between pairs of variables (such as the activities of two structures). As a matter of fact, these coefficients, measuring the cosine of the angle formed by two vectors, i. e. their scalar product, indicate the degree of deformation of the space spanned by the two variables, and represent, therefore, a measure of the forces which act on the structures and induce the correlation. These coefficients, consequently, have the same physical significance as the metric coefficients, which exactly define the measures of the cells when the Gaussian space has "flat" Euclidean properties, while they are only a local approximation when the space is curved. Like the ensemble of the metric coefficients, the set of correlation coefficients represents the fields of force existing in the space, each being a measure of the intensity of the force in a region of that space. Accepting that the correlation coefficients somehow correspond to the metric coefficients, and the latter being the components of the metric tensor, then the correlation coefficients should also be accepted, at a first approximation, as the components of a metric tensor of the cerebral space.

Lest the characteristics of the physical and of the cerebral spaces remain simply at the level of an interesting but formal analogy, we ought also to find the physiological significance of the correlation coefficients to coincide with the physical significance of the metric coefficients. The correlation coefficients, assumed to be components of the metric tensor in the space of the brain functions, besides reppresenting the fields of force in that space, are an actual measure of the degree of interdependence of the nervous structures, inasmuch as they are a measure of the covariation of the variables. This can be said also of the components of the metric tensor in physical space. Being the forces proportional to the amount of matter and energy present, the ten components of this tensor define the extent to which the bodies are reciprocally bound, i. e. the covariation of their respective motions in the generated gravitational field. In the space of the nervous system, the exact nature of the forces can be further discussed, but, as we have seen, their ultimate significance could be attributed in a figurative sense to "attraction", inasmuch as they establish covariation and interdependence of the activities of different structures, operated by the forces reciprocally exerted upon each other. Therefore, there is good reason for considering covariation in the brain space to be homologous to gravitationally induced interdependence in the physical space.

Nevertheless, however close the correspondence between the basic features of the metric tensor in the brain and in physical spaces, and between the significance of the forces acting in the two instances, we must not forget that there is still a fundamental difference preventing the two pictures from being considered fully equivalent. The correlation between nervous structures establishes only a statistical interdependence, which may acquire a non statistical, cause and effect character only for what pertains exclusively to the cross-connections between two structures (thus between the related variables). This was the main reason for seeking in Chap. 8 a definition of the interdependence of the variables which had no statistical character. In the general case, the correlation coefficient is only stating a value of the "coincidental" variations, however produced. On the contrary, gravitation always establishes a true deterministic dependence, and the metric coefficients of the tensor have no probabilistic significance.

But, even if we could not fully accept that the laws governing the deformation of the physical space by physical stresses (such as gravitation) are the same, in the general mechanisms and in their interacting properties, as those responsible for the deformation of the cerebral space (by forces inducing covariation of the activities), it is clear that we neurophysiologists would nonetheless have found a unitary set of principles which determine brain function. At the same time, one has acquired a unified method of investigation: the definition of the components of the metric tensor in the largest possible extent of the nervous circuitry, and within the largest range of different situations. Certainly, it will not be an easy task, actually requiring a great deal of ingenuity to set up the recording and monitor conditions, and suitable computer software for calculating the tensor's components and following their time changes.

## 10.4 Physiological Forces and Tensors in the Central Nervous System

The question of neural forces has been touched upon several times in the course of the preceding chapters (5, 6, and 8). The forces, and a number of factors having the general significance of conductances, define the activity of each structure. However, the nature of the force functions is not homogeneous. Besides those strictly related to nerve impulses, we still cannot precisely qualify those not strictly or necessarily bound to impulses, represented by the so-called modulator molecules, which still have an unclear position in the nervous organization. Moreover, the forces representable as functions of nerve impulses seem to be endowed with quite different properties in the two types of synapses, excitatory and inhibitory. Coupling the concept of force to that of excitation is very easy to accept, but inhibition, from a physiological standpoint, would not seem to be the exact mirror image of excitation, since it is better conceived of as an increased resistance to excitation rather than as some form of added energy. At a closer inspection, however, the impulses impinging on both types of the synapses are seen to act by changing the membrane's conductance to intra- and extracellular ions, thus the forces are not different in nature, but their effects are oppositely directed. Force gradients in the cerebrum are created by the differentiated distribution of the amount of connections, and thus are strictly related to the dynamics of the impulses coursing through these connections and impinging upon the structures, but the effects are measured through the different conductance values, which define at every moment the excitability of the neurons within a structure: the result is anisotropy.

We must figure out how the tensors which induce state changes operate on this structural, chemical and functional substrate. There are several questions needing an answer, the first being the identification of the brain structures which perform the tensor operation (by changing the fields of the forces, i. e. by transforming their configuration into others). Two classes of events at the input of the neurons always contribute to the generation of impulse discharges and of membrane potential oscillations (in a word, to the activity): the impulses in the so-called specific and nonspecific connections. This distinction, however, based on old data and borrowed from the particular situation of the cerebral cortex, cannot maintain its original strict significance when applied generally. Therefore, we shall use it for our purposes only to indicate two arbitrary classes of afferent fibers: i) those coming from the nonspecific or diffuse projecting systems (such as the brain reticular formation, the nonspecific thalamic nuclei, etc.), and ii) those of a specific class that includes all the other types of afferents, belonging to both direct and circuitous connections between the structures. The distinction permits us to identify at least the two main classes of forces having a varying share in the production of general activity, and of output impulse discharges in particular, according to the structure, the time and the state. Both "intrinsic" forces, exerted through specific afferents, and "common" forces, exerted through nonspecific inputs, are always present simultaneously. The common forces are those we have previously indicated to be responsible for the changes of physiological state, such as arousal, REM and non-REM sleep, etc. They account for the generalized changes of the average activity level of nervous structures during the transition from one state to another, by producing "accelerations" in neuronal activity, thus raised or depressed to new steady state levels which, in turn, represent "quasi uniform" motions. Common forces represent not only the operator of the change of level of the nervous activity within each brain subsystem, but, simultaneously, also the operator of the adaptation of the local metrics in the corresponding regions of the Gaussian space, according to the state.

The intrinsic forces act on much more restricted areas of the brain, particularly when the connections between two structures are direct, but the point is that, in these areas, their effects may not be very different from those of the common forces, unless their impulse sequences are patterned by specific stimuli. The common forces are thus the operators of the state changes, setting the stage for the intrinsic forces that find the backdrop of activity and the basic correlations between the variables to be suitable for their action. This situation is not to be regarded as being stable and irrevocably defined, because the common and the intrinsic forces seem to create complicated and everchanging non-linear effects when they occur simultaneously, i. e. practically always, (although they maintain characteristic properties in each physiological state). For instance, in strictly controlled experimental situations, the effects of a stimulus arriving through specific afferents (thus of an intrinsic force), are generally different depending upon the state of the brain (which depends upon the common forces). This type of interference sometimes appears as an "all or none" effect, involving the same efficiency of transmission of specific impulses through central synapses.

Of course, some non-linearity must be expected in many cases, since the ongoing, unstructured activity is only a quasi-uniform motion resulting from a series of transient accelerations; the energizing functions ascribed to mass activity and the reading and gating functions as well (see Chap. 5) are non-linear operations. However, the main source of nonlinearity is the strong interaction between the two types of force. In other words, the effects of the forces on the physiological behavior of the neurons, the deformation of brain space, (itself dependent upon the state of the brain), are rendered non-linear by the reciprocal interaction between the deformation and the forces producing it.

Since the forces are exerted by time-dependent processes, they continually change the components of the metric tensor in proportion to their intensity. The forces transforming the components of the metric tensor (which can be considered a strain tensor) at a time $t_1$ into those at a time $t_2$, form a time-dependent operator, the components of a stress tensor causing the deformation of the space. On account of its anisotropy and non-homogeneity, there ought to be regions in the brain space where the force fields are very weak (the corresponding off-diagonal components of the metric tensor tend to zero), and others where the forces are strong (and, thus, the components of the tensor increasingly different from zero). Therefore any change of the strain tensor, as defined by the metric coefficients, corresponds to changes of both the intrinsic and common forces originated in the same structures. Therefore changes of the fields of these forces correspond to transformations of the stress tensor. The latter, in turn, transforms the strain tensor, and so on. It is obvious that, because of their interactions, common and intrinsic force effects are only distinguishable for the first few moments.

In the matrix of the tensor components, thus, it ought to be possible to isolate regions within which the components' behavior is more or less homogeneous. The situation, again, would never be a stationary one, since a brain state is continuously evolving. However, these fairly homogeneous regions in the matrix (which correspond to anatomically defined structures and circuits) can be preserved if the state is somehow kept stationary. It would be the first practical way to reduce the dimensionality of the brain space, and would represent the physical basis for every analytical investigation of the properties and organization of a subsystem – the only correct way to functionally identify a subsystem, rather than one based only on anatomical connections.

# 11. Outlines of a Theory

## 11.1 The Physiological Fundaments

At this point there is already enough matter for reflection, to sketch a theory, based on principles from different domains of physics, to which the laws governing our brain should correspond. We shall now endeavour to tie the loose ends left in each of the preceding chapters together, and integrate them into as comprehensive, consistent and simple a structure as possible. Nevertheless, what I call a theory is expounded here only as a set of principles that, although logical, are not yet within a rigid framework, but are united only by their internal coherence. These principles are:

1) We do not need the enormous amount of anatomical and functional description of our brain, that is easily found in textbooks. These details are necessary when particular solutions to specific problems are sought, but otherwise tend to impede broad understanding of the fundaments of nervous function. Thus, what remain to be considered are the more general characteristics of the brain as a physical object.

2) The first structural characteristic is certainly the dimensions of the system. The number of neurons ranges in the ten thousand millions, but the overall number of degrees of freedom is, of course, larger. We must not think of fluids in order to find comparable (though, much larger) system dimensions, but of cosmic objects, such as galaxies, which might correspond better to some of the properties of the nervous system. The order that cannot be found, except on a statistical level for some thermodynamic variable, in a volume of indistinguishable fluid molecules is certainly found in the myriads of stars in a galaxy, and even more outstandingly in the brain. The harmonious motion of stars, regulated by gravitational laws, is mirrored in the brain by the different perfected methods of functionally linking the neurons, through an enormously intricate maze of reciprocal connections, and through an extensive system of selective chemical properties. As the stars and their planets compose systems with internal order (true organization) by means of the attractive power of their gravitational fields, so anatomical connections and chemical properties identify, within the whole brain, subsystems and classes of neurons endowed with basic, similar functional properties. However, the coexistence of a stable internal organization based on fine, detailed anatomical connections and chemical properties, with an organization of neurons on purely functional grounds, in terms of heterogeneous classes and subsystems, is peculiar to the nervous system.

3) The high degree of interconnection via direct and indirect pathways between the different parts of the brain, entails that no part can actually work

independently from the others. It is thus impossible to assign an exclusive function to each anatomical subdivision. Because every brain structure is continuously active, at every moment and in every situation all the structures participate, to a greater or lesser degree, in every function. With this, the concept of our nervous system as an anisotropic continuum is beginning to take shape.

4) What we call the function of a part – a neuronal cluster or subsystem – is merely the particular role (eventually also a predominant one) of those neurons in larger operations involving other clusters and subsystems. Each operation requires recruitment of different clusters and subsystems. This mode of operation, in which many parts concur and their individual actions are integrated at some superior level, is "cooperation". Each part cooperates in performing functions, in such a way that the functions do not result simply from the sum of the individual efforts (if there was a way of measuring them in isolation), but from the interactions of the parts (maintained by their reciprocal relationships) which, in turn, ultimately entails a nonlinearity of effects. To reiterate the point, the brain acts as a whole, and the operation of the whole, at every level, exhibits properties not present in the operations of the individual constituents. Consequently, caution is required when appreciating the effects of lesions in brain tissue, since each lesion does not represent a linear subtraction, but damage to the extended system of feedback operations which ensure great interdependence between all the nervous activites. The anatomical connections and the chemical properties (which help define the anisotropy of the system) maintain physical unity within the brain. The functional unity, instead, depends upon "organization", which controls the mechanisms of cooperation through the differential recruitment of subsystems and classes of neurons.

5) The temporal evolution of the whole, which can be equated to the motion of a continuum, is the evolution of its states, the term "state" meaning the set of actual, simultaneous values of all the variables involved in the definition of the continuum. Thus, we define the state of the whole system as a function of all the variables involved. It is obvious that the myriad of infinitely-close successive states cannot all correspond to physiologically recognizable situations. What we call a physiological state thus spans a much longer time interval than that covered by a single set of measures over the whole phase space, which should, therefore, be more aptly called an "elementary" state. Consequently, the physiological state cannot be regarded as a fixed situation, like a frozen section of the brain, but as a time series of continuously developing events which make each instant of the brain's activity (its "life") different from the preceding and the subsequent instants. For the same reason, no single elementary state can be taken as representing a physiological state; only in the case of very gross definitions, such as, for instance, generic sleep and arousal, are elementary states in a long sequence expected to exhibit some common and constant characteristics, albeit restricted to a very small number of variables.

6) Because every brain structure is involved to a different extent in each physiological state, the values of the variable functions for each structure must be expected to continuously change, so that these values will cover a continuum in the phase space. Within the continuum, the peak values will shift over different neuron clusters when the elementary states develop, and each variable's function

will exhibit different values in successive sets of measures, a continuum of values also existing in time (which might lead to true ergodicity, i. e. to the identity of the phase and time averages).

7) Wide cross-connectivity, the continuous activity of every neuron, and the differentiated participation of every subsystem in every function, all point to the conclusion that the fundamental functional property "organization" provides for the distribution of the peaks (the quantity) of activity over the different neuronal clusters. From the different patterns of time and spatial distribution of the quantities of activity finally result the qualitative differences which characterize the different functions and the physiological states.

## 11.2 Steady States and Transformations

Always keeping to the physical, operational side of the problem (for other standpoints see the Appendix), the organization of brain functions and of the state of the whole can be viewed both in steady state conditions and time-dependent evolution.

8) Statistical mechanics permits us to follow the temporal development of the elementary states under time-independent (actually truly stationary) conditions, by considering each of them as a point image moving in a space with twice as many dimensions as there are variables. Thus, the position of the image point in the space is defined by an equal number of parameters. Since the stationary condition corresponds to that of a closed system, the image point would belong to a surface of constant energy, representing a physiological state, (each physiological state, thus, on a different surface), and the transition from one physiological state to another would require a "jump" of energy. If energy is not added, the image points of the elementary states are "trapped" on one of these surfaces, which together they define with their continuous wanderings, covering almost the whole surface with their everchanging positions, (i. e. if sufficient time is allowed, attaining ergodicity). According to this, the states can be discriminated only on the basis of their energy content, a nebulous concept in the context of the nervous system, bearing only a statistical connotation. As a matter of fact, an average measure of energy – the most important phase measure of the state of a system – has a meaning only if the system is sufficiently homogeneous and isotropic. The usefulness of energy measurements would then be restricted to isolated subsystems or classes, in which homogeneity and isotropy can at least be assumed.

9) When the system is considered to be made up of a great many similar non-interacting subsystems – a Gibbs ensemble – all in their natural (time-independent) development, then, by the Liouville theorem, there is conservation of density (of the subsystems) in phase space. If we add a high degree of interactions between the subsystems, (as is found in brain ensembles), this must lead to maintenance, not only of the density of the subsystems, but also of their mutual relations. If the natural motion of the ensembles in the brain can be assumed to correspond to a steady state (not a truly stationary) condition, then, since the steady state implies constancy of activity, and thus of the interactions between

different subsystems, there will also be a constancy of relations between the subsystems (maintaining, in this way, a non-statistical density in phase). This assures that measurement of average energy can also be performed in non-isotropic systems such as our brain, provided steady state, if not truly stationary, conditions are met.

10) Measurement of energy in the nervous system as measurement of a state of motion becomes possible through the Lagrangian and the Hamiltonian functions based on the generalized coordinates of position and velocity (or momentum). With the latter functions, however, the measurement of energy must no longer be statistical, since the dynamic variables (the physiological parameters) that define the motion of the system define its state in terms of kinetic energy. Another non-statistical definition of state in the nervous system is obtained through the set of Onsager's phenomenological relations, in which suitably selected nervous variables represent Onsager's fluxes and forces: the matrix of the Onsager coefficients relating the fluxes and forces describes the state.

11) The generalized coordinates of Lagrange can be applied to the nervous system to define two types of motion: i) constant velocity (uniform, stationary, or inertial) motion, which neurophysiologists call spontaneous, or, more correctly, ongoing (steady state) activity; ii) accelerated motion, such as that induced by a force, which neurophysiologists call induced, or evoked activity (a product of stimulation). The usefulness of this mechanical approach is that through the generalized coordinates of position, velocity, and momentum we can define, besides energy, "generalized" forces in the continuum. The brain forces, although of different natures, can be reduced to a common term, a function of the trains of nerve impulses, which constitute the main representation of nervous activity. Only suitable functions of the impulses can be translated into generalized coordinates to fit the formalism of the generalized forces.

12) Although the differential spatial distribution of anatomical interconnections creates an inherent field heterogeneity, by no means is it the unique or the most important source of differentiation within the field. The "structure" of the continuum is not defined by the fixed irrevocable system of connections, but by an everchanging pattern of lines of force, created by a time-dependent functional anisotropy, substituting for the fixed anatomical pattern. Probably only in conditions of extremely depressed brain activity may the two patterns, anatomical and dynamic, coincide. It is apparent that what we now call the everchanging pattern of lines of force corresponds to the differentiation within the endless sequence of elementary states of our nervous system.

13) The main consequence of the standpoint that total brain activity must be viewed as a sequence of states, is the focusing of our attention on the laws of transition ("transformation") from one state to another, which constitute the real subject of interest. The task of individually identifying each elementary state of an endless succession is obviously a hopeless one, besides probably being physiologically irrelevant. In principle, the goal can be reached by different routes. Quantum mechanical treatment suggests the possibility of defining physical energies in the nervous system, through generalized coordinate representation. The computation of matrices resulting from transformation of the generalized coordinates and momenta, and of other matrix functions of the generalized coordinates,

ought to produce the energy states of the system, and the probabilities of transition from one energy level to another, i. e. from one state to another.

14) The transformations of a nervous system continuum in terms of the quantum mechanical concept of state vectors seems to be most fruitful. By considering the "agents" of the transformations, the approach may better account for the evolution of the system. All the variables of nervous activity can be represented as vectors. The state of the system, a function of all its variables, (a function of these vectors, then), is a vector itself, the "ket" vector. As in quantum mechanics, with values unified in arbitrary and normalized units, the discriminant factor is no longer the ket vector's module, but only its orientation in space. The contribution of each vector in defining the direction of the state vector is determined by a coefficient, essentially the "weight" of each vector in that state. The quantum mechanics principle of superposed states is directly applicable to brain function, although somewhat relaxed, since different components comprise any nervous subsystem or class. The ket vector (the state) can be represented by the set of the coefficients of the component vectors, i.e, by the pattern of quantitative relationships between them. The elementary nervous state is then represented by the instantaneous pattern of spatial values of the coefficients (a phase pattern). A first approximation of the pattern might be the relative orientation in space of pairs of variables defined by the cross correlation coefficients. The set of ket vectors derived from the temporal development of the elementary states, represents a physiological state.

15) Vector representation, although having a different aim from quantum mechanics, is not inconsistent with our picture of the nervous system as a continuum, or as fields of forces. Transformation of the fields of force (transformation of the continuum) is best represented as transformation of Gaussian coordinates by linear operators, the tensors. The metric tensor is the most useful general tensor for our purposes. The matrix of its components, which are expressed as partial differentials of the elements of the space, represents the actual pattern of relationships between the variables constituting the coordinates of that space. Each matrix component defines the "local" participation in the overall space deformation that is induced by forces which change the relationships between pairs of variables. Essentially, both the coefficients of the vector representation and the coefficients (or components) of the metric tensor define the pattern of mutual relationships between the elements of the space, which constitutes the state of the system.

16) In terms of the forces involved, thermodynamics also represents the distribution of relationships, through the whole set of the Onsager phenomenological relations between the variables involved. The Onsager equations, although called phenomenological, state a deterministic cause and effect relationship between a driving force $X$ and its effect $J$ (a flux) in terms of a coefficient $L$. The matrix of the coefficients (straight, or diagonal, and mixed, or off-diagonal) between all possible pairs of forces and fluxes exactly defines the state of the system, by defining, in each region of the space, the share of a given force (the strength of its field) in determining an effect. The difference between this matrix and the matrices of the correlation and of the metric coefficients, lies in the fact that in the Onsager set the cause and effect (force-flux) relationships

are based on precisely defined "affinities" and are subject to precise constraints. In the nervous system these affinities can be represented only by known anatomical relationships (fiber connections). This precludes the immediate utilization of phenomenological equations over wide areas of the brain, because the functional and anatomical connections, as well as their quantitative aspects, are known only to a limited extent and in circumscribed subsystems. On the other hand, for this very reason, the method is perfectly suitable for analytical research.

17) A physical description of the evolution of states of activity presents other possible pictures, assuming the states to be vectors (the ket vectors) and the dynamic variables to be operators on these kets. The Schrödinger solution leads to the "wave" equation, applied to nervous variables and states, in which the time-independent operator is the total energy of the system. The Heisenberg picture presents the reverse situation, the operators being time-dependent and the states fixed. The situation in the brain seems to be better portrayed by the Heisenberg picture: its time-varying operators anatomically and physiologically correspond to those subsystems with diffuse projections which normally induce changes of physiological state, and are most probably also responsible for the endless sequence of transitions between the elementary states as well. In this way, the transitions could be considered the result of "unitary" (or "contact") transformations. Finally, the transformation functions of the representatives of the state ket (which can be viewed as a coordinate system) would allow us to compute the probabilities of transition.

18) The theory of perturbations of a system can account for the changes occurring in the state of a system in steady state conditions (if considered as a natural motion). Perturbations may induce very small changes within quasi-stationary motion, measurable as oscillations of energy content, not large enough to induce the energy jumps necessary for sudden changes of physiological state, but occasionally sufficient for elementary state transformations.

19) The endless transformation of states is more comprehensively solved by the time-dependent tensor operation. Besides describing the temporal evolution of the complex system, the Gaussian tensor operator also accounts also for quantitative invariance, through "covariation" of the variables, determining, in turn, the invariance of the associated functions. In terms of the continuum, the stress (or force) tensor and the strain (or deformation) tensor in the nervous system are interdependent. The stress tensor changes of the strains (the values of the relationships between variables) within the field; in turn, modification of the strain tensor changes the forces, i. e. the stress tensor, on account of the widespread interactions between structures, again in an endless succession. The tensor representation of the phenomena has the advantage of conceptually unifying all the state transformations, of the elementary as well as the physiological states, in the same way.

## 11.3 Methods

Brain "functions" can be studied through different methods: i) the activity of single neurons, classes, or subsystems; ii) the production of localized lesions in different structures in order to detect their share in the performance of the whole; iii) behavior, i. e. outward manifestation of nervous activity. These apparently different methods of investigation can plainly be unified by the common principle of establishing, first, the pattern of relationships between the widest number of brain variables, then, of observing the temporal evolution of these phase patterns (what we have called the transformation of elementary states).

20) Nervous activity and its corresponding outward manifestations must both be quantified in terms of the same patterns of relationships. Of course, every method of recording nervous activity has its own methods of quantification, but the different methods essentially refer to two main types of activity, that of single cells and that of large populations. Since comparison of absolute values has little or no significance, only normalized values are allowed. A physiological function implies a brain activity of a more or less extended duration, so sequences of such normalized values must be considered over a long period of time.

21) Since a physiological function cannot be described by a single variable (that is, a single neuron, class, or subsystem), but by the ensemble of the variables, what counts is the pattern of distribution of the activities in the phase space and the temporal pattern of their development.

22) Averages over very short intervals can describe the time course of activity of brain variables, either of single cells or of populations, because various brain structures may exhibit the same time patterns for up to several minutes. However, short-duration averages has another great advantage, is the "language" of the brain, comprising short-lasting (milliseconds) nerve signals, understandable only in groups, phase patterns, and time sequences. Averages with suitable time constants (which constitute a critical parameter for each physiological situation) are a way of detecting modulation of the signals representing nervous activity. Other functions of the activity may better describe the operations of nervous structures, but the short-time averages are sufficient to support the assumption that the whole brain's operation can be represented through a comparison of their relative time courses.

23) Another aspect of the averages is statistical. Many neurophysiologists still believe that multidimensionality and quantitative variability require that the system be treated as a stochastic source of signals. From this viewpoint, the only reasonable strategy is again based on averages and comparisons of their values. These facts point to the same conclusion, our fundamental basic principle that brain activity can be studied and described only within some characteristic time period.

24) Comparison must aknowledge that at different times the same neuron clusters are involved in somewhat different brain functions. Therefore, what is significant for each nervous function is the temporal pattern of correlation of the activity of every structure with each of the others, obtained by computing the time-series correlations between the structures, for the period necessary to describe the physiological situation.

25) The other feature of the unified methodology is the definition of the correlation pattern. The statistical correlation coefficient (as the Pearson product moment) representing the cosine of the angle between a pair of vectors, measures the slant of the coordinates in Gaussian space. Through these coordinates, the

metrics of the Gaussian space can be exactly defined (if the space is Euclidean) or locally approximated (if the space is curved). On the other hand, the metrics of the cells in the mesh of a Gaussian space is more aptly represented by metric coefficients, the components of the metric tensor. Of course, the statistical measure (the correlation coefficients) is based on averages, so that, again, the characteristic time for the average correlation must be determined empirically, and will be, probably, differ according to the variables and physiological situation. Nonstatistical correlation methods can only be based on partial differentials of pairs of variables (were this performed with average measures, a statistical connotation would always be present). However, since regular correlation patterns appear only when the correlations are computed within some characteristic time, their physiological significance could be linked to the existence of a time characteristic for each brain function. The latter point, however, requires more direct experimental support.

## 11.4 Perspectives

The conclusion of the preceding section questions the order of problems which can be investigated according to the principles of our theory. In the next chapter we shall see how the theory is naturally applied in investigating some fundamental brain problems, such as "personal" time, the "covariation" of the activity of different structures, the operation of nervous forces, the invariance of functions, the conscious "observers" within the brain. These problems will lead us from purely neurophysiological themes to speculations nearer to philosophy, but the very possibility of widely spanning the field of neuroscience makes me feel that the principles of our theory, though briefly outlined and still incomplete, are sufficiently general for approaching problems of different complexity and at different levels, which was my original purpose.

26) In Chap. 4, we discussed the analytical method, keeping it distinct from the method adopted in the theory just outlined. It should now be plain that the differences are principally of mental approach, and are reduced to minor details about the practical execution of research, so the procedures typical of our theory are perfectly suited for analytical investigation. The two procedures may chiefly diverge as to the use of data. By trying to analyze the structural anatomical details of a circuit, and the behavior of the neurons creating it, as a function of some particular stimulus, one merely isolates a subsystem, and works with it by establishing its initial conditions and its subsequent changes of state. The whole experimental process can always be viewed as the assessment of the transformations that a tensor, or another operator induce on the state of the circuit's elements. Of course, the analytical approach should investigate the simultaneous relationships between all the variables of the circuit, i. e. the multidimensional approach of the covariations.

27) The theory predicts that following damage to tissue the function of the remaining structure will not undergo a simple deficiency syndrome, i. e. a mere quantitative defect, but that there will be a qualitative change of the functions performed. The remaining structures may have been deprived of necessary infor-

mation, or the functions may have been taken over by other structures as part of their own functions (as plasticity assumes). In all cases, a different distribution of the lines of force (a different pattern of lines) in the remaining tissue, is likely to cause new phenomena to appear and others to disappear. Investigating the nature of these phenomena, however, will require the preparation of accurate tests for their detection and measurement.

28) It goes without saying that pharmacological and neuropsychological investigations can be conducted in exactly the same way as physiological ones, with the great advantage that data from whichever line of research will be immediately comparable, being expressed in the same, common terms. It will not be a minor achievement, even if it should remain the sole one.

## 11.5 Gathering the Threads

I have no difficulty in admitting that the points listed in the previous four sections do not yet resemble a well-defined theory. As a matter of fact, in Points 1 to 7 I have proposed the "fundaments", which are merely a set of conclusions drawn from an established body of knowledge. Some of the propositions are, certainly, widely acceptable; some, however, having been filtered through a personal interpretation of the facts, may be debatable, particularly the central issue, the participation of all the brain subsystems in every cerebral function.

It is all a matter of perspective. The investigator engaged in analytical research seeks, by his very approach, to define the details of a specific function within anatomical circuitry. He chiefly tries to outline it by tracing in it the finer connections. In this way, the specific function is studied in isolation, i. e. paying little, or no, heed to what is happening, at the same time, to other functions in distant or adjacent circuits, which are, however, connected to the former either anatomically or by chemical affinity. However, there is no doubt that every circuit is always active, and that current refined tracing methods are continually discovering new, unsuspected connections which steadily make the spatial range of influence of any neuronal pool appear wider and wider, well beyond the classical projection fields described in even recent books. Of course, the amount of participation of the different circuits in each function is the discriminant factor, and it actually defines the differences between one function and another, between one state and another state, within the context of the principle of the "whole operation". From this follows, quite naturally, the necessity to consider the total temporal evolution of the subsystems as a unity, subject to "a principle of organization" by which different subsystems simultaneously change their activity according to the function being performed, thus, to the state of the whole. The same concept could be stated in a minimally different, but much more emphatic form – different subsystems simultaneously change their activity according to the function "to be" performed, thus, to the states "through which the whole must pass". However, these two different points of view are largely presented in the Appendix.

The description of the system "brain" under steady state conditions and in temporal evolution may be obtained using different procedures. Points 8–19 contain propositions about how the whole can be viewed by building a continuum. What are proposed are not models, as I have insisted throughout, but suitably differentiated approaches to truly different aspects of the same problem. The state of the continuum with its fields may be described through the Onsager's matrix of the relations between the forces and their effects in the maze of the circuits, or through the matrix of the crosscorrelation coefficients. The first is characterized by cause and effect relationships, the second by purely statistical correlations, albeit not altogether divested of deterministic implications.

However, the temporal evolution of the system requires something more. The quantum mechanical concept of state vectors and operators may be applied to the single "state vector" ("state function", or "wave function") according to the Schrödinger picture of a time-dependent state and a fixed operator, or the Heisenberg picture of a time-dependent operator and a fixed state. Setting up and solving the appropriate equations might, however, prove to be an unapproachable task. On the other hand, representing the state of the whole in matrix form and the temporal evolution of the system as transformations of the matrices in time, offers three well-defined, complementary methods. The matrices of Heisenberg, Born, and Jordan allow us to deal with discrete states and their relative probabilities of transition, while the matrices of the (statistical) crosscorrelation coefficients and of the metric coefficients (as components of a metric tensor) are, perhaps, more easily submitted to time-dependent tensor operations which transform them into others, and may also account for the invariance of the functions. The practical application of these different methods to the experimental data will certainly bring to light unforeseen difficulties, but, again, we adhere to the principle of a common, general approach based on the full set of interrelationships.

However, Points 8–19 are not simply propositions for suitably representing brain functions in terms of physical principles and related mathematical procedures; they stem from hypotheses, or physiological "constructions" advanced from Chap. 6 onwards: i) the Gaussian characteristics of the space of brain functions and states; ii) the nature of the variant and invariant functions and their relationships within the Gaussian space; iii) the manifold aspects of "time" as a factor characterizing each function, and thereby affecting the values of the corresponding variables; iv) "local" and "personal" time, both state-dependent and state-defining, as involved with the other dimensions in the deformation of the space by neural forces (the "covariation" of time with the other space dimensions); v) the tensor nature of the operators of transformations within the brain continuum. All these constructions give rise to actual predictions, which may or may not be confirmed, but their internal consistency makes a real theory from the 25 points, from the fundaments to their practical physical representations. The predictions are logically contained in the nature of the system and the physical context in which it is considered, and are not the product of personal interpretations. However, if the usefulness of a theory is the advancement of verifiable predictions, only the fulfillment of the latter transforms a theory into a scientific achievement. Experience will provide the final verdict.

The theory has taken a definite (though by no means necessarily definitive) shape in a relativistic frame, something which is quite unusual in neurophysiology, and quite unexpected when I set out to organise my ideas on the general principles of cerebral function. There is the relativistic transformation of time, which seems to "covary" with the functions and, therefore, with the state of the brain; there are the state-dependent and function-dependent changes of the metrics of the Gaussian space. Of course, it is too early and much too hazardous to foresee a "geometrization" of the nervous system, like that which in Einstein's relativity was produced for the physical world. However, we can easily foresee presently unimaginable difficulties in this hitherto-unexplored experimental approach; therefore, without going any further with the theory, I shall try to outline, in the next chapter, a few of the problems which may arise once the relativistic frame is accepted.

# 12. On this Side of the Border: Relativistic Aspects

## 12.1 Is There a Brain Relativity?

Let us now presume that the principles outlined in the preceding chapter are convincing and acceptable enough to provide a reasonable basis for future experimental testing, and that the many related technical problems, which need to be solved somehow, are already behind us. The purpose of this chapter, therefore, is not to add other principles, but to pursue some of the lines of speculation that stem directly from our sketchy theory, as its logical consequences, and to outline the problems involved.

The conclusions that nervous activity is best represented in a general Gaussian space, and that the multitude of relationships constituting its states and their evolution is best investigated through the tensor approach, do not themselves suggest a discrete branch of physics which may have the same order of problems as our nervous system. The physics of continua, which we have accepted as a suitable framework for the brain's functions, offers a variety of choices, including the mechanics of rigid and deformable bodies, elasticity, fluid dynamics, etc., but no one of them alone would fully represent the fundamental problems of our brain if not also considered within a relativistic context.

With the thorough analysis of space, time, and forces, relativity (both special and general) has profoundly revised the philosophies of our Universe since the Renaissance. Analysis of the brain's properties, and their possible description through known physical laws, has naturally led us to outline a picture of cerebral function which adopts several principles of relativity, a theory endowed with profound philosophical implications. This encouraging result provides a "philosophy" of the nervous system, a set of principles of theoretical neurophysiology. It has been a stimulating task throughout, not leading, of course, to any final discovery, but simply offering further evidence that an object which "moves" in a domain can do so only in obedience to general principles, whether the object is a lump of rock or a sophisticated, complex, and baffling thinking machine, like our brain.

Some references to relativity have been made, but, essentially, relativity is relevant to the nervous system for two reasons: i) brain subsystems exhibit activities that can be classified as uniform (inertial, time-independent), or accelerated (time-dependent and force-dependent) motions; ii) brain forces may induce a change of reference system (for instance, by "curving" the cerebral space time, as gravitational forces do with physical space time).

The crucial point thus seems to be the presence or absence of forces which may accelerate the motion (the brain activities), and curve the space coordinates. However, if we do not think (or cannot demonstrate) that, at any given moment, time is affected, it would be altogether superfluous to consider our problems within the context of relativity (either special or general). The real, discriminant factor is time. Events within the CNS (relative motions and effects of forces) obviously cannot affect physical time (the time of our clock), but can affect "personal" time. However, we ought to demonstrate that we are justified when we substitute personal time for clock time.

## 12.2 Time

This subject has been discussed, repeatedly and at length, by scientists, philosophers and theologians. Feeling uneasy about the subtleties of a philosophical discussion, my purpose is not to forward new propositions about the nature of physical and cerebral time, but to gain some insight into Man's subjective measurement of time. There is no direct objective evidence for the existence of personal time. It is still only a hypothesis, although a very likely one, based upon the judgement of time intervals, which have been demonstrated to vary in different brain states. Should further experience confirm that the subjective judgement of time varies appreciably and systematically with every change of state, personal time itself would acquire some reality of its own, supporting the hypothesis that, in the brain space, it is endowed with the same properties and significance that clock time has in physical space. In the following discussion we shall assume this reality.

The philosophical significance of time is not separable from that of space. Time is measured as an interval between two events, but the ultimate or primary measurement is spatial, such as (in earlier epochs) the star's motion through the sky, or (nowadays) the space covered by light in vacuum. While space is measured with spatial comparisons (to objects of unit length), the legitimacy of a time interval measurement, by whatever clock it is performed, comes from its relationship to an interval of space. Clock is a man-made device; in nature, there are only events, sequences or intervals which may be artificially divided (either by measuring the displacement of a stick's shadow, the amount of sand falling in an hourglass, the number of oscillations of a piezoelectric crystal, or the emission of atoms in matter).

Conceptually, time and clocks are so tightly connected that whenever physiologists mention time in relation to brain-controlled functions, they always look for a neural clock, but, in principle, there is no reason why there ought to be one in the brain, nor has there been discovered so far such a temporally regular succession of neural events that could reliably be utilized for time interval measurements. We can only be trained to divide the intervals between internal or external events, either consciously or instinctively guessing the length of an interval by comparing it to some "stored" unit of length (if there could be one).

If we assume, as there is good reason to suspect, that personal time in the brain space behaves like physical time in physical space, it would seem illogical to measure them in different units. We should, therefore, use for both the unit "$ct$" (the space covered in the time interval $t$ at the velocity of light $c$). We should ask ourselves whether this is actually necessary, since measurement of $ct$ could pose serious problems (such as the absence of true displacements in nervous "motion", motion of very low velocity, etc.). However, on account of the likelihood of relativistic phenomena, such as the personal time dilation that presently concerns us, assuming the unit $ct$ as a measure of time would seem advisable.

## 12.3 The Velocity "c" in the Nervous System

The conclusion of the preceding section requires further discussion. In physical, four-dimensional Minkowski space, the velocities of macroscopic bodies are generally much lower than that of light; therefore, with the Lorentz relativistic correction, the deviation of the time measured by a clock at rest from that measured by an accelerated clock is infinitesimal. Only for extremely large (cosmic) spaces and velocities does the relativistic formula reveal measurable differences. However, in the brain space, things could be different, having to deal with a space that is not measurable with physical objects, and the observed events not being displacements. In the brain, unlike physical space, we never have to deal with light, or phenomena at velocities comparable to that of light, so the use of the velocity of light as a measuring unit does not seem justified. On the other hand, the coordinates of the brain space are changed (perhaps, curved) by the forces within the brain, and the personal time might be dilated. Hence, the $ct$ parameter in the brain space (if it has to be retained) should consist of $t$, corresponding to personal time, and "$c$", however the latter not definable as the velocity of light.

Let us see what the properties of the parameter $c$ should be in the nervous system. In relativity, the velocity $c$ represents a limit, a velocity that cannot be exceeded, and thus can be taken as a term of comparison for all other velocities "$v$" measured in physical space. However, the absolute value of $c$ would not have real intrinsic importance, since the relativistic formula for time dilation depends upon the ratio of the two velocities, $v$ and $c$, squared not upon their absolute values. If, instead of the value $3 \times 10^8$ m/s, the velocity $c$ were much lower, say in the order of $3 \times 10^3$ m/s, it would be in the same range as the velocities of present day vehicles, and the latter would become relativistic velocities with corresponding measurable deviations of the time parameter. (Of course, besides such a down-scaling of the reference velocity, we should also assume the state of matter to be compatible with these values; $c$ being an electromagnetic unit, all properties of matter would be greatly affected at these lower velocities.) In a system like the brain, not moving in physical space, where time is assumed to be personal, not physical (clock) time, the value of the reference velocity $c_{CNS}$ can be different from that of light, provided it has the same characteristics, and fulfills the same function, as $c$ in physical space.

The value of $c_{CNS}$ can be calculated, as in special relativity, through the Lorentz transformation formula, using the experimentally measured deviation of personal time from clock time, with the value of $v$ known (see the next section), having been experimentally determined. This value of $c_{CNS}$ might correspond (or be proportional) to the maximal theoretical "velocity" of a nervous event – some known phenomenon associated with nerve impulses or their functions – although we have not yet demonstrated it (selecting the suitable variable will probably require a lengthy procedure of trial and error). After all, even in physical space, the velocity of light has a limiting value only in regions free of matter (where gravitational forces are absent) i.e. in flat, Euclidean space, where light can be propagated in a straight line. The corresponding situation in the brain space would be represented by the absence of a difference between personal time and clock time, entailing the absence of forces (intrinsic and common), which would imply that the brain had scarce, or no activity (profound coma or death) and there would be no nervous events to measure the velocity of. In all other instances – the presence of forces, as considered in relativity – the value of $c_{\mathrm{CNS}}$ may differ from the absolute limit, depending only upon the ratio of the space coordinates to time.

In the general case, $c_{\mathrm{CNS}}$ must be defined experimentally, but not as the maximal velocity of nerve impulse conduction in the fastest fibers. That would be a measure of a fallacious $c_{\mathrm{CNS}}$, since it would be completely unaffected by relativistic transformation of the brain space (i.e. independent of the known nervous forces) and, therefore, it would not have the same character in the brain space as $c$ in physical space. The consequence is that the value of $c_{\mathrm{CNS}}$ can only be calculated, as in general relativity, from the ratio of the value of some other suitably selected brain parameter to the personal time, that is, relative to the forces present, as is the case for $c$ in physical space.

### 12.3.1 The Velocities "$v$" in the Nervous System

As it was said before, the velocity $c_{\mathrm{CNS}}$ must represent the limit for a velocity $v_{\mathrm{CNS}}$, not yet defined. The brain space being a space of functions, the velocity $v_{\mathrm{CNS}}$ cannot denote a velocity related to a displacement in physical space (as the velocity $c_{\mathrm{CNS}}$ is not the velocity of light) and cannot be related to the so-called velocity of conduction of the impulses from one neuron to another, since it could not be influenced by those forces which affect the brain space time. Thus, we are constrained to look for other neural velocities; a likely candidate might be, again, the parameter selected in Chap. 8, the rate at which the impulses impinge upon the synapses, which is merely the rate of generation of the impulses in the neurons of origin, subject, as such, to all the brain forces previously identified. For those variables not directly expressed in terms of nerve impulses – such as, for instance, the brain waves- the velocity could be represented by the first time derivative of their instantaneous values, in whatever units are appropriate. (However, there being doubts about their self-standing identity, we should avoid this procedure.) If we take $v_{\mathrm{CNS}}$ to be equal, say, to the time rate of impulses impinging upon the synapses, then $c_{\mathrm{CNS}}$ could well represent the upper limit of this velocity, thus acquiring, finally, that well-defined significance we were not able to make clear before. Now, assuming this significance for $c_{\mathrm{CNS}}$, it is easy to see that the limit of the total velocities exists in the brain space as well as in physical space (although restricted to the same nerve ending) since the time interval between successive impulses can never be shorter than the duration of a single nerve impulse, and two impulses cannot be superimposed, i.e. cannot be present at the same time in the same point. This shows that a limiting velocity $c_{CNS}$ can be experimentally determined independently of the time dilation formula as in special relativity, taking heed of the fact that the limit can be reached only in the absence of restraining (dissipative, delaying or inhibitory) forces. Commonly, the highest observed impulse rates do not exceed 300/s when the restraining forces are assumed to be absent, say in epileptic discharges, (although there have been occasional reports up to 500/s). In the vast majority of situations the restraining forces are present to a different degree, and in each of these situations the experimentally computed $c_{CNS}$ (and $v_{CNS}$) must be correspondingly lower. As we shall see next, to the increase in inhibitory, dissipative, or negative modulator effects corresponds a decrease in impulse frequency, i.e. a lengthening of the impulse intervals, in our model, a decrease of velocity.

### 12.3.2 Mass, Velocity and Energy in the Central Nervous System

The relationships found in relativistic motions between mass and energy are not exactly mirrored in the central nervous system. In the latter we cannot refer to the mass of an impulse, since we can conceive the latter only as being similar to massless particles (see Sect. 8.2), therefore, the related impulse energy would be proportional to the time rate of the impulses in a train, which has the significance of a frequency. As a matter of fact, the energy could be computed as $h\nu$ ($h$ = Planck's constant, $\nu$ the frequency, or rate). Since the neurons are continuously subjected to forces which change their frequency of generation, we expect, in place of the impossible relativistic effect on resting mass, relativistic effects on the frequency of the impulses, actually a decrease of their frequency (like the red shift of the radiation from a gravitating source). By treating the emission of impulses as a radiation, the similarities of the two phenomena seem to be pushed a bit too far, however striking they might seem. Nonetheless, each neuron can be thought to receive (could we say "absorb"?) a radiation $h\nu$, that is, a train of impulses (like a "wave packet"?) at some frequency, and to emit a radiation $h\nu$ (another impulse train) at another, generally lower, frequency (on account of the necessary spatial and temporal summation of the impulses at the synapses). In every case, the frequency of the impulses in the output train is always also influenced by the inhibitory forces that decrease the impulse frequency, like radiation emitted by an increased mass (the relativistic effect). Remember that synaptic inhibition, by raising the resistance of the neuronal membrane to depolarizing influences, can be considered as if it were increasing an inertial mass (which would exert a stronger attraction upon the outwardly directed radiation). It is a widespread, though never clearly expressed, belief that the decrease of the frequency of the outgoing impulse discharge may also be ascribed to dissipative processes within the neuron, (a cause not excluded, although not of a demonstrable nature). While such a mechanism could partially account for the basically lower impulse output frequency, it surely cannot represent a principal, or a verified way of controlling the neuron's output (unless we may attribute the latter function to the unknown mechanisms of modulator molecules). However, at present, the membrane resistance seems to depend mainly upon inhibitory influences.

The progressive decline in the frequency of the output discharge, an actual frequency modulation of the output train of impulses, is also consistent with the delayed, or progressively increased effect of forces. On one hand, the recent developments of research on the chemical modulation of synaptic mechanisms seem to support the hypothesis that synaptic mediators slowly change their effects on the postsynaptic membrane, as the molecule is progressively metabolized. On the other hand, the inhibitory, long, and short feedback circuits bring about a delayed, progressive decrease of the output discharge from a neuron, proportional to its initial frequency of emission. In all cases, forces which can affect the space of the brain functions can produce, as a temporally delayed consequence, a decrease of the frequency of discharge. This temporal separation of events is possible in the nervous system on account of the slowly-developing effects of mediator molecules and the low velocities of impulse propagation in nerve pathways, but has no analogous counterpart in the inanimate, physical world.

## 12.4 Inertial and Accelerated Systems in the Brain

It has been seen previously, that personal time gives rise to subjective judgements of shorter or longer time intervals (as compared with clock time). These time-interval judgements can reveal appreciable relativistic deviations when treated with the Lorentz transformation, as we have hinted in the preceding section. In our everyday experience in physical space the Lorentz-Fitzgerald effects of space contraction and time dilation, are infinitesimal, since we normally deal with velocities far below the velocity of light. However, as I have maintained, in the brain space things might be different. In the brain, the velocity $v_{CNS}$ (the rate of the impulses) and the corresponding limit velocity $c_{CNS}$ (if we consider the impulse frequency) are never too different. A ratio $v : c$ of 0.2 is common, while in many experimentally verified situations a ratio of more than 0.8 is observed. This means that in almost every case the ratios of the squared velocities we have to deal with in the CNS would be classified as relativistic in physical space. (It must be remarked that high frequencies are never maintained for more than fractions of a second under physiological conditions, a negative modulation of the impulse rate intervening rapidly, as discussed in the preceding section.) On the other hand, it is conceivable that the factor involved in determining the extent of time dilation might not be the impulse rate directly measured as such, but a function of it, or that its values might be subjected to gradation (scaling) within the nervous system, justifying the appearance of relativistic effects even at lower impulse rates. Therefore, pending further neurophysiological experimental support, the question of the mechanisms of time dilation will not be pushed further.

If we assume that the velocities within the nervous system can be such as to give appreciable deviations of the personal from the clock time, then, a number of physiological problems arise. First, it ought to be considered that the personal time is a variable of state (since, ultimately, both $v_{CNS}$ and $c_{CNS}$ are variables of state) which covaries with the other variables, as was the hypothesis of Chap. 9. However, we cannot expect that all the innumerable variables of the brain will covary in the same way in each state. Some will probably remain fairly constant or changing only temporarily, others will exhibit covariation of different degrees. As a matter of fact, by computing the cross-correlation coefficients between the variables, it has been shown that in each of the fundamental brain states (REM, non-REM sleep and arousal) the average covariances of different pairs of variable are different. Of course, the heterogeneous distribution of forces among the nervous structures is responsible for the different degrees of covariation, so that different structures (and different functions) are differentially affected in a state change. This appears to be a crucial point, since the presence of both covarying and non-covarying variables means that there exist subsystems which are subjected (at least temporarily) to forces, and others which are not. In other words, there must be subsystems in uniform (time invariant, stationary, or inertial) motion, and subsystems accelerated, thus, respectively, with a flat, Euclidean space geometry, and with a curved one. What we have called personal time should behave accordingly in each subsystem, i. e. it should covary with both types, remaining unaffected in the subsystems in uniform motion, while undergoing dilation in the accelerated ones.

We see that this conclusion leads us, unexpectedly, to admit the simultaneous existence of more than one personal time in the same brain. It is a rather surprising conclusion, although perfectly logical, forcing us to necessarily assume the presence, in the brain, of the typical relativistic situation, that is, of systems in different types of motion, thus, in the same brain, of potentially different "observers", one for each subsystem. The problem seems so important that we should now more closely investigate the possible multiplicity of personal times. In Chap. 9 we have been led to infer the existence of a personal time by the observation that time interval measurements vary with the state of the brain (at least, they have been shown to be different in two characteristically different states), exhibiting dilation in a state that could be compared to an accelerated motion. This is not an autonomous and independent "measure" of time in the brain, but a personal, subjective impression (experienced as a surprise), appreciable only if referred to a clock time "at rest" in the physical system outside the brain.

In relativistic situations, the time dilation has two aspects: an accelerated observer receives the signals emitted by a source in inertial motion at longer intervals, while, for an observer at rest with the source, the clock of the accelerated system is slowed down. For a large number of systems, all in different types of relative motion, there must be an equal number of "local" times, and this must be exactly the situation for the brain as well. Then, the problem is to find out what relates all the different local times to each other, since there must be a common basis for all the time dilations, and there is no reason to assume that any one of the systems is a reference for all the others, all being in the same relative situation. This has been a main concern of relativity theory. The basis of time measurements in relativistic problems is a quantity, the "proper" time $d\tau$ (the "Eigenzeit"), defined only differentially and not integrable. It is related to what Einstein called the "natural volume element" dx1, dx2, dx3, dx4 (x4 representing the time coordinate in the Minkowski space), a function, of space and time.

As a matter of fact, the quantity $d\tau$ is defined by the space coordinates as well as by $ct$, therefore, by the actual "world line" of the moving system in four dimensional space time, depending upon the path followed and the ratio of the system's velocity to the velocity of light. As a consequence, since each system follows its own world line, each system in motion has its own proper time, which remains invariant as long as the system is in uniform motion, but is subjected to change whenever an acceleration is produced by forces. Thus, each brain subsystem in motion relative to the others, must have its own time, which becomes smaller the larger the ratio of its own velocity $v_{CNS}$ to the limit reference velocity $c_{CNS}$. It is, actually, through the proper time concept that the famous "clock paradox" or "twins paradox" was explained, since the world lines of different systems can be compared only on the basis of their associated proper times.

Being a function of space time, the proper time is a function of the metric (fundamental) tensor in Gaussian space, i. e. of all the coordinates of this space, where the forces affect the space time, accordingly making the proper time covary. The set of coordinates similarly changed by the forces may be considered as comoving coordinates for the brain functions performed within the space spanned by those coordinates; the same can be said of the proper time, which represents the time in a system of comoving coordinates.

Before we may tackle the problem of observers within the brain, let us consider another aspect of the ratio of the squared velocities $v$ and $c$. This ratio determines the extent of the time dilation, but (simultaneously) also of the space contraction (the Lorentz-Fitzgerald contraction), time and space covarying in a reciprocal way. It is not easily perceived what a contraction of the brain space parameters actually means. Obviously, no hints are available in neurophysiology (the problem has never been raised), but it is worthwhile to push the formal reasoning to its logical end. In previously reported experiments, where it could be assumed that a time dilation (the deviation of personal time from clock time) was present, it could be further assumed that a space contraction occurred as well. In that situation, the experimental conditions were characterized by enhanced brain activity, i. e. the brain was kept busy at some higher level by demanding tasks. Since the brain space is a space of functions, the space contraction, if present, either did not impair the performance connected to the functions, or even improved it. It must be remarked that in different brain states (which should correspond to different degrees both of shrinking of the space coordinates and of time dilation) the brain does not perform the same functions in exactly the same way. Each function is performed somewhat "differently" in different states, though remaining basically the same function (i. e. retaining an invariance). The differences are appreciated through descriptive terms, like more or less precision, increase or decrease of completeness or of involvement of other functions, increased or decreased decision time, and the like, all terms having, at the same time, a quantitative and a qualitative connotation. There is no way yet of venturing hypotheses on the significance of space contraction in physical and in physiological sense, until more experimental data is available and tests of function performance have been developed. We can only guess that the increased involvement of subsystems, as required when the brain is engaged in demanding tasks, results in an increased correlation between the variables involved in the different elementary functions connected to the task, with a corresponding tendency to colinearity of the coordinate axes. As a consequence, the volume of space spanned by the axes would be reduced, and this might represent a type of space contraction, perhaps mirroring, in physical significance, the effects of contraction of the physical space coordinates. Pushing the analogy further, we may recognize in the contraction of the physical space as well an increased cross-correlation between the coordinates, since in their "global" or "local" contraction there is covariance, being the changes of the measures of the coordinates defined by the metric tensor, and related to a common quantity, the natural volume element, therefore related with each other and with time in a unique way.

### 12.4.1 The "Observers" in the Brain

The simultaneous presence within the brain of subsystems in uniform and in accelerated motion, raises other problems. In every complex physical system, the motion of each subsystem is either uniform or accelerated, and thus there can be simultaneous and independent observers for each of the subsystems, who are living with their respective proper times, and can simultaneously and independently report the time dilation and space shrinkage they observe in the others. In the brain, to admit the simultaneous presence of more than one observer is awkard. Although each subsystem, as we have seen, has its own proper time, every person can behave according to only one personal time and can report only one "impression" about his own time (the difference between his own and the clock's time). So, there must be a hitherto-unknown link between the subsystems' proper times and the unique personal time. The latter, for example, could well result from an average of all the proper times, or represent only one proper time, for instance, that of the structure (or structures) from whence our consciousness arises, or toward which attention is directed. The uncertainty rests not in the nature or the significance of proper time, which is a physical quantity, but in the way it influences our personal judgement (or sensation, or feeling) of time. But we have to leave the problem unsolved.

Looking at the question from other standpoints, we could ask ourselves whether a uniformly moving subsystem or an accelerated one may be the actual observer within our brain. We can put this in a somewhat different way, i. e. whether there can be in the brain a fixed, absolute reference (which is impossible in physical space). In physical systems, time dilation and space contraction can be observed, for instance, by an observer at rest in one system who sees the events developing in another system in the same or in a different type of motion, and measures them with his own rod and his clock. An observer at rest in any system has absolutely no way of being aware of his own space contraction and time dilation, since his space and time measuring devices are affected in exactly the same way as the space time in which he is at rest. An accelerated observer can feel the forces acting on himself, for which he has receptors in his body, but his measuring rod and his clock cannot give him any clue. Though, again, in the brain the situation could be more complex. If the observer could have a look into his own brain, at closely succeeding time intervals, the same observer could be aware of time dilation and space contraction of his own subsystems, and report about it. Is this an impossible situation? There have been numerous reports of "feelings" of time actually "slowing" (see also Sect. 9.3.2), but the situations were not controlled enough to allow appreciation of the data in relativistic terms. Anyway, the presence of an internal "time reference" in the brain does not yet seem to have been demonstrated in carefully designed experiments, so that we still feel entitled to conclude that the brain observer cannot be split between the two types of subsystems, but can be at rest only in one, and is unable to observe the others, and must refer his experience (judgements or feelings) of time interval length to systems outside the brain.

In the case of invariance of functions, judgements, etc., we may also ask whether the observer (the subsystem) performing the judgement ought be at rest in the subsystems in which the measurement is being made. This would seem a necessary and unavoidable requirement, since, if the measurement is to be correct, there ought to be neither time dilation nor space shrinkage relative to each other, otherwise the observer would be in a position to see them (which, apparently, is not the case). This requirement can be met only when the relationships between the subsystems are preserved by covariation, which is when the observer is at rest in the measured subsystem. However, we have in the brain an organization by which some of the functions are "located" ("dominant") on one side of the cerebral hemispheres (the cortical part of them) and others on the other side. In the previous example, the recognition of faces "belongs" to the right hemisphere, whereas verbal and numerical symbolization and, therefore, interval measurements belong to the left one. The first situation refers to a function retaining invariance in widely varying conditions, the second one seems largely state-dependent. The latter is believed to be so (see Chap. 9) because it needs reference to external measuring standards, clocks and rods, which cannot be made to covary by the brain forces, (only personal time and internal "memorized" rods, if ever possible, could covary, then revealing their lack of congruence with outside standards). Is this, perhaps alternately, a matter of observers which cannot be brought to rest in the particular subsystems? There is no necessary difference between the two explanations, since to bring an observer to rest in a subsystem means, actually, to make the observer covary with the subsystem, so the problem is always one of covariation. Only carefully planned experiments will produce the correct answers, further speculation would be futile.

# Appendix

## Beyond the Border: Metaphysics and the Brain
## A Sample of Problems

### A. 1 Introducing the Problems

In this Appendix we shall try to look beyond the fence, trespassing as neurophysiologists across the border to metaphysics and extending to the latter the ideas expounded in the preceding chapters. I will not propose a new metaphysical theory or a systematic examination of the field. Rather, I will try to bring closer together two opposing views which, by the particular natures of the system we have been investigating and of the theory proposed, could be reconciled, at least in part. It will not be a learned discussion. On one hand, the coexistence in the brain of structural and physicochemical, but also of cognitive, generally psychological, ethical and spiritual problems, challenges the neurophysiologist to find answers on the basis of organization, notwithstanding the limitations of his knowledge. On the other hand, the possibility is offered to him of confronting his conclusions, made on purely neurophysiological and physical grounds, with those reached from opposite standpoints by non-physiologists, philosophers and theologians. However, I shall only refer here to what has been written centuries ago by Thomas Aquinas, whom we may consider a most representative scientist (and also a philosopher and a theologian) of the Middle Ages, who elaborated Aristotle's science and, by fusing it with theology, built a "corpus" of doctrine with which our confrontation must begin. Going through his writings (Summa Theologiae and others) as well as through a few modern commentaries upon his thought, was initially only leisure reading, but my attention was soon fully captured by the unfaded freshness of his arguments, the depth of his logic, and the far reaching truth of many conclusions, maintained until the present day. However, I do not feel sufficiently competent to provide the extensive, penetrating and exhaustive discussion the matter deserves, so what I am going to do is, essentially, nothing more than point out the themes where Aquinas and neurophysiology seem to meet, a more extensive comment requiring much deeper knowledge of both sciences.

The problems pertaining to this Appendix are wide-ranging. They represent a constant source of debate for everybody involved in the investigation of the phenomena of Nature, because they reach to the core of the very principles of scientific research itself. Perhaps, too few investigators are aware of the fact that the way they approach research, the relationships between Science and Nature, is primarily the expression of their personal way of conceiving Nature and the essence of the phenomena constituting its continuous "movement".

Reduced to its simplest terms, the question involves the two main, opposed views about the world in which we live (including the whole Universe). The modern scientist of whatever domain of Science, reaching farther and deeper every day with his powerful means of investigation, pictures to himself the world of Nature as being self-consistent and self-sufficient, a world in which all the phenomena derive one from another in a most necessary sequence, determined by natural laws, i. e. by the extended system of relationships which bind together the elements of matter, from the macro- to the ultra-microscopic levels. It is an utterly positivistic and empirical standpoint, adherence to which is justified by the fact that experience has demonstrated it to be perfectly suited for investigating all the natural phenomena, provided the inquiry is not pushed to extreme limits, such as the ultimate structure of matter, the extension of the Universe, its beginnings and its ultimate fate. Within these limits, everything can be explained without any reference to external or primary causes, or to other factors not already accounted for by the interplay of physicochemical forces. The success of the positivistic, empirical approach has convinced most students of Nature of the actual uselessness of taking into consideration the other view, which claims independence from the physicochemical deterministic chain of events leading to the former picture of the world, and adopts a more subjective and autonomous standpoint, a "spontanéité spéculative" (to echo D. Dubarle). From his independent position, the scientist strives to recognize the fundamental principles which the system as a whole obeys with its extended multitude of relationships, and which account, as well, for phenomena escaping empirical, analytical investigation. So, we have, fundamentally, two philosophical views. On one hand, that of a self-sufficient Nature, whose movements (actual motions, changes, developments) have in Nature itself their own moving agents, the determinants of the endless chain of cause and effect relations – a Nature self-created and self-organized through the mere quantitative ratios of its components, which define the forces operating in it (though with a beginning which is still veiled, and an end which is shrouded in utter uncertainty). On the other hand, a transcendent, initial, primary cause and its chain of effects, a set of principles which determine the development of the physical world from beginning to end, a finalistic motion. These two views have too often and too superficially been tagged, respectively, as modern and medieval natural philosophies. Modern scientists, apparently, have reason not to bother much about these problems, they just note their existence and some of their consequences to idly speculate upon them, but nothing more. Modern research has developed enough intellectual and material tools to investigate the development of material events, in their logical consequences, their mathematical exactitude and predictability. As I said previously, at the borderline, where modern tools are unfit for proceeding further, there being no natural development to follow, no sufficient data to base predictions upon, no preceding experience to take as a model, there, at the borders and beyond, lie our problems.

I have never thought the two points of view necessarily contrasted greatly with each other. Certainly, everyday experience in the laboratory permits us to explain natural phenomena without recourse to external factors, or reference to primary causes: for each event can be described the preceding conditions, the forces involved, and the future developments. In this respect, Nature is certainly self-consistent and self-sufficient. However, this is not the point. Even in a purely mechanistic approach, we must be aware of the fact that Science aims to uncover the relationships between both components of Nature, the material and the non-material. This means that our knowledge, our science, heavily depends upon the way our brain sees Nature, and, in turn, that the relationships we are uncovering are those "we" can understand. If we accept this proposition, there is no reason why we should attribute scientific status only to concepts which refer to what is in some way measurable, and not to others. We can accept Einstein's standpoint that hypotheses which cannot be tested should be dismissed as useless or altogether untrue, but only in so far as we can assume that everything existing in Nature ought to be exactly measurable. But Man's brain, besides the concept of a measurable Nature, according to which physical phenomena of whatever essence can be quantified, has other concepts concerning less easily quantified categories, like good, evil, ethics, which do not have yet a distinct, universally accepted, standard measuring "rod", but are, nonetheless, measurable by each individual with a measuring standard of his own. In its essence, this is a situation to which the physicists are accustomed, that of different systems in motion relative to each others each one following its own "world line", with a local geometry and a local time, i. e. with its own measuring standards. Famous scientists all over the world have been disturbed, at first, by these relativistic principles; however, these are now largely accepted, because they still provide the most logic and comprehensive framework for physical events, even if the experimental tests are far from easy and not free from errors. The category of concepts such as good, evil, etc., and those related to physical world are still considered as the two opposite poles of our brain's products. But, "natural" philosophy, i. e. speculation about the essence of every physical phenomenon of our universe, has no sharp borders towards metaphysics, towards the philosophy of the essence of immaterials, of what transcends and escapes our measuring instruments, of the "esse" in itself. Natural philosophy and metaphysics are like a Janus Bifrons, two faces looking in opposite directions but belonging to the same head, using the same speculative power of a shared brain: the truths discovered by each pair of eyes are fused together and contribute to the same comprehensive knowledge. The positivistic, empirical approach and free speculation should not be strictly confined, each into its natural field, respectively, of measurable physics and immateriality. Both approaches have principles and concepts profoundly rooted in logic, and inasmuch as they represent a "rationale", they must have a common, innermost root. Thus, we are perfectly justified when we tackle both types of problems with the same intellectual tools, provided we are aware of the peculiarities of the two opposite situations. The borderline between physics and metaphysics is still a no-man's-land, where a natural scientist may probe the ground with his concepts, to find out how far he may proceed in unveiling the detailed machinery of Nature. However, he is not justified when he shuns problems that are not mechanical, because at these borders he may be able to unveil the principles, the logic that guides the organization of the mechanism, perhaps, the very purposes of its existence. However, for this aim, he must claim his right to intellectual freedom, the right to also approach the natural physical questions the other way around, with that "spontanéité spéculative" which does not proceed step by step along the infinite chain of cause and effect relationships, but directly probes the possible logical frame in which both material and immaterial things have their own place. This approach is not like taking shortcuts, the procedures might be even more elaborate and painstaking than in the other approach, but certainly bring more complete and all-embracing results.

Thomas, the medieval philosopher, has no difficulties accepting the positivistic, empirical standpoint; he too recognizes that in the physical world there is a fatal chain of events, determined, however, by a primary (not itself determined), matter-independent cause. He also maintains that between non-living phenomena and life there is a difference, but of a quality which does not interfere with the "mechanics" of the physicochemical phenomena, but is concerned with the finality the latter subserve and, therefore, resides in the transcendentally logical principles of the living machine. In a beautiful work on Thomas's philosophical and theological thought, A.-D. Sertillanges comments on his standpoint with these words: "Pour lui, dans l'organisme, tout se passe chimiquement comme s'il n'y avait point d'âme, et tout se passe vitalement comme s'il n'y avait point de chimie". The problem is seen, thus, at two different levels, but the core of that part of Thomas's doctrine in which we are interested will help us to understand how these two levels of natural events may create a unity.

As far as the brain is concerned, what is urging a neurophysiologist to assume unusual standpoints is not the desire to make a question of faith in this or that philosophy; simply that, when considering a system like our brain, the need for different standpoints arises spontaneously. In physical, inanimate Nature, as I said previously, everything is explicable up to the limits where, the usual philosophy of research failing, solutions have so far been sought on the basis of chance (as at the level of the ultrastructure of matter) or arbitrarily (as at the level of some constants in cosmological formulas). In the brain, the relationships between the components are not as clearly stated as in the inanimate world; although I have made a tenet of considering the nervous system as any other physical system, it is now apparent that it also possesses its own peculiarities, besides life. A mathematical predictability (within limits) like that of the physicochemical relations is also found in the brain, but only, and to a very limited extent, for some circumscribed, partial functions, such as, for instance, reflexes. Outside these simple instances, the lack of predictability and the presence of phenomena (ideas, behaviour, etc.) which are not immediately accountable for by the known laws of Nature, make us think that we are always confronted with problems "beyond the limits". It is obvious that, at this stage, much of our present uncertainty about the brain's manifestations must be attributed to the still extremely scarce knowledge of the extensive relationships between its components, but it is equally obvious that the innermost principles and causes of its structural formation and of its functional organization are far beyond the reach of any empirical, analytical approach, since they deal with another category of phenomena. Unfortunately, these are problems upon which the neuroscientists are encroaching in their everyday activity, therefore, it is absolutely necessary that, apart from strictly intrinsic mechanistic cause and effect relations, we also purposefully search in our brain for the general principles, the general plan, the causes of determination, the key to understanding its beginning and its future development. In Max Born's remark that theoretical physics is actually philosophy, although not expressed, is implicit the belief that the investigator must have a philosophical standpoint to start with (a belief he actually translated in the quantum physics principles, a real natural philosophy); the same conviction was often clearly expressed by Albert Einstein too, albeit he reached quite opposite philosophical conclusions. Neither of the two great physicists could possibly be considered a medieval scientist, both are founders of the most modern, pregnant and fruitful theories about the essence and mechanics of the physical world. In the following few considerations I will strive to maintain as neurophysiological an approach as possible, aiming, however, to clarify where an actual philosophical standpoint, in Max Born's sense, may lead us when we confront metaphysics, starting from our few theoretical, neurophysiological principles.

The lively natural philosophy of antiquity which could explain the phenomena of Nature only on the basis of the extremely scant scientific information available, and was still capable of surprisingly far-reaching conclusions, died out in the late Middle Ages after Aristotle's science had been revived and fused with the theology of Albertus Magnus and Thomas Aquinas. From the latter stem the clearest concepts about the ties binding Nature, Man and God. The fading of interest in natural philosophy was conditioned by the very completeness of the picture reached, by the consistency of a vision which left, for that time, very little to further improve or broaden, and, there being no new science to be confronted with, the way was open to the most abstruse and (for us today) mainly futile, discussions among the scholastics. Only in the following centuries, with positivistic science acquiring an increasingly independent position, could Man acquire, through it, a steadily increasing self-assurance, developing new instruments to explore further into the reaches of the Cosmos and deeper into the structure of matter, revealing more matter and its related physical and chemical laws where only faith dominated previously. So, Man arrived at loosening the bonds between Science and God, as they were conceived from Antiquity to the Middle Ages, (the action of the Church against Galileo was a last outcry of the old, dying conceptions).

As I said, I would like to recommence a comparison of our scientific thought with that of the greatest of the medieval philosophers, Thomas Aquinas, because, in the search for the theoretical bases of neurophysiology, I found strikingly close reminescences with principles developed in his many natural and theological scripts. The procedure I shall be following is the same as in the preceding chapters. As we have investigated the correspondence between physical and brain laws, so we shall examine here the brain's laws through the screen of Thomas's thought, trying to understand how close to his principles are the neurophysiological principles previously described. I shall keep an agnostic position while tackling problems involving religious beliefs, because the purpose of this Appendix is not to affirm the truth of this or that standpoint, but to show that what faded out in the Middle Ages must be considered as simply forgotten, and that the problems are still there, awaiting the contemporary natural philosophers and scientist-theologians to recommence the search for their solution.

## A. 2 Man and Animals

The basis of what I am going to discuss, is the existence of a clear-cut neurophysiological distinction between the intellectual "powers" of Man and of all the other living beings. Man is continuously using his superior intellectual power for submitting to his necessities not only the whole inanimate world around, but all the living beings as well, by conditioning their growth and their very existence to his welfare. This is no feat of brute force; through his intellect, Man invests the instruments he has created with the mechanical force needed for his purposes, such as the use of the concentrated power of atoms for different tasks. The most "intelligent" of the apes is barely able to conceive by itself of using a stick as an extension of its arms.

We have seen (Chap. 3) that the main feature enabling Man to exhibit superior intellectual performance is the development in his brain of structures and organizations through which functions like abstraction, symbolization, and thought are performed. The direct consequence of this fact is the larger predictability of the animal's reactions to external stimuli when compared to the human ones. Strangely, this predictability has fostered a widely extended neurophysiological investigation with animal brains, considered as useful "models" of the human one, which is, instead, unpredictable to the greatest degree. In animal sensation, the paired sensations of pleasure and displeasure seems to dominate, determining corresponding behaviors of approach-avoidance, attack, escape, even of fierceness and tameness. On the contrary, the freedom displayed by Man from sensations and from external events is amazing; he can neglect and sacrifice himself for the common welfare, can program the future, can concentrate on spiritual thoughts, can suffer unbearable evils. The largest component by far of Man's behavior is certainly not in direct relationship to external events: different philosophies and ethics guide his motions more than contingent situations, although, surely, many of the immediate causes of the motion may come from without. So, when generally talking of "superior" intellectual powers, I do not imply that the quantity of intellect of a Man is greater than that of a chimpanzee. The difference is "qualitative", in that Man has a larger neuronal mass, yes, but such as to provide for different structural and organizational features in his brain, which certainly enable him to also perform quantitatively better than apes, but substantially enable him to do different things, as well to do the same things differently. Whatever might be the origin of this differentiation, slow evolution or step jumps as by mutational changes (the latter hypothesis seems to better fit the qualitative aspects) the result is that the appearance of the human brain represented something like a crisis in the history of the Universe and, even if extended in time, perhaps, through humanoid steps of development, comparable to the crisis that brought out the formation of the cosmos (the "Big-Bang"), a unique, unrepeatable and, probably, irreversible event. Extensive behavioral and ethological investigations have failed to detect any continuity in the "qualities" of the intellectual performances of present day beings, so that the gap between the highest primates and Man still justifies the term "crisis".

The distinction I am maintaining exists between the intellectual power of Man and of brutes, was certainly more decisively expressed by the scientists of previous centuries. Thomas repeatedly returns to this point in several scripts, by keeping a distinction between intellectual and sensitive "souls", the former peculiar to Man, the second to animals: "Homo intelligit quia principium intellectivum est forma ipsius" (Sum.Theol. Ia, q.76, a.1) (see Note 1) and again "... est enim animal per animam sensitivam, homo vero per animam intellectivam" (Sum.Theol. Ia, q.76, a.3) (see Note 2). I cannot explain here the inexact correspondence of the Latin term "anima" with the english term "soul": as long as non-religious problems are concerned, I will retain the term anima. So, according to the preceding quotations, Man only has intellectual power, whereas animals can only "sense". The significance of this distinction is that animals are "inferior" because they cannot reach knowledge of the truth, which is only attained through rationality : "... alia animalia sunt infra hominem, quod non possunt attingere ad cognoscendam veritatem, quam ratio inquirit." (Sum.Theol. Ia, q.79, a.8, ad 3m) (see Note 3). From the senses, in fact, the animal cannot expect a true image of the universe: "Naturas autem sensibilium qualitatum cognoscere non est sensus, sed intellectus" (Sum.Theol. Ia, q.78, a.3) (see Note 4), and "... non est expectanda sinceritas veritatis a corporis sensibus" (Sum.Theol. Ia, q.84, a.6, quoting Augustine) (see Note 5). I think we can adhere without hesitation to these sentences, because reaching knowledge of the nature of things means being able to abstract from the external qualities, i. e. from what fills the senses, the common properties constituting the idea of the thing, a real process of analysis and synthesis that only the rational power of the adult human brain can accomplish. Therefore, it is obvious that since what the senses can tell the brain represents only the appearance of the contingent, no one, animal or Man, can expect from them the truth, which refers instead to the immanent, the essence, the "esse" of things. There is also in Thomas's sentence a surprising anticipation of modern neurophysiological findings about the not absolute reliability of nervous messages from the sense organs as faithful representations of the external qualities of the world. Central innervation of the sense organs, by allowing interference by centrally originated activity in that of sensory structures, is steadily changing what the senses tell our brain. This also justifies on objective experimental grounds the conclusion drawn by Thomas: "... imaginem corporis non corpus in spiritu, sed ipse spiritus in seipso facit" (Sum.Theol. Ia, q.84, a.6) (see Note 6), which stresses the true character of the human intellect, its independence from the contingencies of nature, i. e. the independence of Man himself, perhaps, hinting at the reversed order of dependence to be realized in the relationships between Nature and Man. In what sense the human intellect is said to be independent, is accurately specified by Thomas in several places, where the relationships between the different brain operations are investigated: "Licet intellectus operatio oriatur a sensu, tamen in re apprehensa per sensum intellectus multa cognoscit quae sensus percipere non potest" (Sum.Theol. Ia, q.78, a.4, ad 4m) (see Note 7), that is, the content of the sensory information is only the occasion for further operations of the brain, which, after taking heed of sensation, proceeds by itself. As a matter of fact, the brain of Man has the peculiar power to use information from the senses to build a different image of the external world, nearer to the essence: "... tamen est quaedam operatio animae in homine quae dividendo et componendo format diversas rerum imagines, etiam quae non sunt a sensibus acceptae " (Sum.Theol. Ia, q.84, a.6) (see Note 8); in other words, Man can see beyond appearances, constructing the truth by himself. I think that these quotations are self-explanatory, once it is accepted that by "dividendo et componendo" is meant that sequence of analysis and synthesis which can be accomplished only by a rational power.

The consequence of all this for the neurophysiologist is that the animal's activity can only be congruent with the messages from the external world: we may say that animals are compelled to act by what their senses transmit to their brain, whereas Man has the possibility of other choices, dictated by his rational operations. In brutes, in fact, the sensory operation can activate only a compulsory attraction or avoidance, whereas in Man these "primitive" reactions can be overcome by other, more complex operations. Neurophysiology and ethology demonstrate that knowledge of the external conditions is sufficient for a reliable prediction of an animal's behavior, but not of a human's. The same conclusion is beautifully stated by Thomas: "In aliis enim animalibus statim ad appetitum concupiscibilis et irascibilis sequitur motus, sicut ovis timens lupum statim fugit: quia non est in eis aliquis superior appetitus qui repugnet. Sed homo non statim movetur secundum appetitum irascibilis et concupiscibilis; sed expectatur imperium voluntatis, quod est appetitus superior." (Sum.Theol. Ia, q.81, a.3) (see Note 9). This sentence is also self-explanatory, at least on theological grounds, but in it we can find matter for further reflection. Here comes, in reality, the problem of human freedom in its full extent. The indirect and non-obligatory relations between human activity and external stimuli was in Antiquity and in the Middle Ages the self-explained consequence of the superiority of the anima intellectiva of Man to the anima sensitiva of animals. The difference resides in the rational powers of the former, his capacity for deliberating, by simultaneously weighing the sensory pulls, the circumstances, the predictable consequences, the finality of the action. There is no doubt that the human brain has these powers, thus, also on neurophysiological grounds we can expect that the decision to yielding to a drive or not is not defined in advance. On the other hand, always on the same grounds, it could be maintained that the large degree of unpredictability of Man's action is the outcome of the extremely widespread connectivity between all the brain structures: through it the final decisions are submitted to a complex of utterly unknown variables, and for an observer the result is, thus, determined by chance, only subjected to some statistical prediction. However, it is easy to see that Man's choice of the response to a drive is never chaotic, but always corresponds to a rationality of judgement, in whatever direction it is made (but more of this will be found in Sect. A. 7).

## Notes

When translating an old language like Latin into a modern one, an extended knowledge of both is required, such as this writer unfortunately does not possess. I have simply endeavoured to give the most approximate significance to several metaphysical concepts, concisely expressed in Latin, but rarely equally concisely in modern terms. I apologize to the reader for translating most sentences literally: the purpose was to preserve as much as possible the flavour of Thomas's phrasing, which is admirably concise in its construction.

*Note 1.* Man has intellectual operations because his formative principle ("forma") is intellectual.

Here, the difficulties of concisely expressing the tightly related philosophical concepts of "forma", "principium" and "anima" (to be encountered next) already become apparent. We shall find that the "forma" is actually the "anima", the principle forming (or characterizing) each being.

*Note 2.* The essence of the animal is the sensitive formative principle (the "anima sensitiva"). The essence of Man, on the contrary, is the intellectual principle (the "anima intellectiva").

*Note 3.* The other animals are inferior to Man because they cannot attain the knowledge of the truth that rationality can investigate.

*Note 4.* Knowledge of the inner nature of the sensory qualities does not pertain to the senses, but to the intellect.

*Note 5.* The sincerity of truth cannot be expected from the body senses.

*Note 6.* The image of the material body is not transformed by the body into the spirit, rather, the spirit transforms the image (of the body) into itself.

*Note 7.* Admittedly, the intellectual operation may originate from sensory experience; however, of what is being apprehended through the senses, the intellect may gain much knowledge that the senses cannot perceive.

*Note 8.* There is in Man some operation of the anima, which, by dividing and reconstituting, forms different images of things, including those which cannot be perceived by the senses.

*Note 9.* In the other animals, motion consistently follows a strong desire ("appetitus") to grab or repulse something. For instance, the wolf-fearing sheep consistently takes to flight, because in animals there is no other superior desire which can prevent it. Man, however, does not move consistently according the desire to grab or repulse, but waits for orders from (his own) will, which is superior desire.

## A. 3 The Intellect

What is, in modern scientific terms, the "intellect" of the medieval philosopher? Thomas says that the intellect is not a substance, but a power of the anima"... virtus intellectiva non est virtus alicuius organi corporalis ... sed in materia est, inquantum ipsa anima cuius est haec virtus, est corporis forma, et terminus generationis humanae." (Sum.Theol. Ia, q.76, a.1, ad 1m.) (see Note 10), and " ... necesse est dicere, secundum praemissa, quod intellectus sit aliqua potentia animae." (Sum.Theol. Ia, q.79, a.1) (see Note 11). Translated into our language, intellectual power is not built into the structure, in the "matter", but resides in what we have called the "organization" of the latter. Every neurophysiologist will concede that organization and structure are interdependent, but are still separated categories; only a trend of modern research maintains that structure alone is the determinant factor, by considering as fundamental only the way the brain elements are materially interconnected. Since in animals the structure of the brain is simpler than in Man, the organization of Man's brain must be, to say the least, at a level higher than that of brutes. If we may say that "higher" has not simply a quantitative connotation, but actually indicates a qualitative jump, then we may also assume that the powers of the anima (not the anima itself, of course!), either sensitive or intellectual, represent the different organizational plans, or rather, the principles which constitute the essence of their organization. I have been unable to find a more appropriate neurophysiological term than "organization", which implies both the "principle" of organization, the purely nonmaterial aspect, and its embodiment in the structure of the brain circuits.

The independence of the intellect from matter has been stressed by Thomas in different places, making clear that the alteration of matter can prevent the intellect only from manifesting itself externally: "Cum intellectus sit vis quaedam non utens corporali organo, nullo modo impediretur in suo actu per laesionem alicuius corporalis organi, si non requireretur ad eius actum actus alicuius potentiae utentis organo corporali." (Sum.Theol. Ia, q.84, a.7) (see Note 12). Here the concept of the human intellect is manifested as exclusively spiritual power, not bound to matter in order to exist, but needing matter only to express itself. So, what is usually meant by "intellect" is the sum of all the intellectual powers residing in our brain, of which all, or a few, may be operating at any given time, but all the same exist as a whole in the brain, as the faithful representatives of the really superior nature of Man.

The interpretation of the powers of the anima as identified by the principles of organization, therefore expressed by the integration of the different structural elements and their functional properties in a "whole" function, actually corresponds directly to the interpretation given by the philosophers and theologians of the early and late Middle Ages. In Thomas's sentence "Forma mixti corporis habet aliquam operationem quae non causatur ex qualitatibus elementaribus" (Sum.Theol. Ia, q.76, a1) (see Note 13), is the philosophical and theological essence of the concept repeatedly sustained throughout the book, that the brain operation is a whole based on both elements structure and organization, since "forma" of the "composite" (mixti corporis) means "anima" ("Id quod est forma dicimus animam", Sum.Theol. Ia, q.76, a.1) (see Note 14), the principle dominating with its powers the life of matter. The union of body and anima, thus, endows the whole complex with properties which are neither of the body nor of the anima alone. This is the concept of a superior unity stemming from the union of different entities (which applies at a lower level as well, to the enhanced performing potency due to cooperation of elements of purely material essence).

The concept of unitary operation of our brain, by which phylogenetically older structures have grown together with newer ones, so that phylogenetically older functions have been modified following their integration with newer (and different) functions, was clearly expounded in the Middle Ages. What I said in Chap. 3, that the basic animal functions are performed by humans in a different and "perfected" way, because older and newer levels of organization have grown and fused together, finds in Thomas's words a precise expression: "Sic igitur anima intellectiva continet in sua virtute quidquid habet anima sensitiva brutorum, et nutritiva plantarum." (Sum.Theol. Ia, q.76, a.3) (see Note 15). The concept is expanded by explaining the practical consequences of this situation, because it is clearly stated that there is actual interaction of lower and higher levels of organization. At every stage of development there is fusion of lower with higher powers, to a different degree for each anima, with the result that the "whole" actually exhibits a continuum of properties, in which each component somehow shares something of its own essence with the others, without, at the same time, losing its own peculiarities, but actually gaining in determination. Again, Thomas: "... quod est inferioris preexistit perfectius in superiori" (Sum.Theol. Ia, q.76, a.5) (see Note 16), and, once more, "Inferiora autem sunt in superioribus eminentiori modo quam in seipsis" (Sum.Theol. Ia, q.84, a.2) (see

Note 17). A further step with the same concept of a continuum of properties is the sentence: "Unde essentia superioris creaturae, etsi habeat quamdam similitudinem inferioris creaturae prout communicant in aliquo genere, non tamen complete habet similitudinem illius, quia determinatur ad aliquam speciem, praeter quam est species inferioris creaturae" (Sum.Theol. Ia, q.84, a.2, ad 3m) (see Note 18). The belief in a continuity of evolutionary, but also mutational, partial changes, is clearly perceived in these sentences, which concern not the material condition, but the essence of the innermost nature of beings, as appearing in the powers of the different animas, and concerns matter only secondarily, through them. Once the correspondence between the powers of the anima and the principles superseding each level of organization is accepted, the truth of the preceding sentences cannot be questioned even on neurophysiological grounds. However, if we accept this, we must proceed in the direction indicated by the medieval philosopher. The fact that the powers of the human anima also include those of lower species, places Man's anima (therefore Man himself) in the highest position of a hierarchical order of living beings. However, it ought to be clear that the hierarchy is not of beings, but of properties, and each being may find a place in it because the anima spans a range of properties, so that the limit of the range for any one represents the beginning for another. Therefore, the hierarchy in this continuum also implies that the lower-placed beings participate in some of the higher properties, a concept already expressed by Dionysius Areopagita and concisely quoted in Thomas's words: "Supremum inferioris naturae attingit id quod est infimum superioris." (Sum.Theol. Ia, q.78, a.2) (see Note 19).

## Notes

*Note 10.* The virtue of intellect does not belong to some body organ ... though it resides in matter, inasmuch as the anima having this virtue is the formative principle of the body, and the end-point of Man's generation.

*Note 11.* According to what was previously said, we must say that the intellect is a power of the anima.

*Note 12.* The intellect being a force not using a material organ, its action cannot be prevented by lesions of any organ of the body, unless for its action is required (in turn) the act of some power making use of a body organ.

*Note 13.* The principle forming the composite has some type of operation not derived from the elementary qualities of the latter.

*Note 14.* Forma is what we call anima.

*Note 15.* Therefore, in this way, the intellectual anima includes among its capacities whatever is possessed by the sensitive anima of brutes and by the nutritive anima of plants.

*Note 16.* Whatever is possessed by the inferior, exists in a more complete way in the superior.

*Note 17.* The properties of inferiors exist in superiors in a more distinguished way.

*Note 18.* The essence of a superior being, albeit somewhat similar to an inferior one, inasmuch as they have something in common, is not, however, completely similar to the latter, bacause it has been defined with the purpose of realizing another idea, besides that of the inferior being.

*Note 19.* The highest virtue of the inferior nature attains the lowest one of the superior nature.

## A. 4 The Continuum

In the closing sentence of the preceding section there is a concept which, if accepted in neurophysiology, becomes one of its fundamental principles. The modern students of Thomas stress it, recognizing its importance in modern philosophy. The point is that in the conception of Dionysius and Thomas, Man (and his anima, and the powers expressed in the principles of organization of his brain) occupies an intermediate position in the order of Nature. Man finds himself between the material and the spiritual domains, actually linking them, since he participates in both, like a bridge between the purely material and the purely spiritual aspects of Nature. Man, in other words, is the essential link in the continuous chain of the components of Nature, not confined within the material domain, since he participates with the sensitive part of his anima in the world of brutes, but connected by the intellectual part to the lower levels of the spiritual domain: "Unde dicendum est quod nulla alia forma substantialis est in homine, nisi sola anima intellectiva: et quod ipsa, sicut virtute continet animam sensitivam et nutritivam, ita virtute continet omnes inferiores formas, et facit ipsa sola quidquid imperfectiores formae in aliis faciunt." (Sum.Theol. Ia, q.76, a.4) (see Note 20); and "Anima autem intellectiva secundum naturae ordinem, infimum gradum in substantiis intellectualibus tenet." (Sum. Theol. Ia, q.76, a.5) (see Note 21).

Let us now see which of these concepts is present in the theory of this book. The first, and essential, as I said previously, is the concept that Thomas has of the continuum, as an ordered sequence of properties of the substance that extends from inanimate matter to include plants (with the anima nutritiva), brutes (with the anima sensitiva) and the supreme living being, Man (with his anima intellectiva) to abut onto the non-material, spiritual world. The continuum of Dionysius and Aquinas has a precise connotation in the theory of this book, since it represents the field of the powers of the anima, identified, in the brain, in the principles of organization necessary to "move" the matter. If we are willing to accept the correspondence between powers of anima and principles of organization, there can be no doubt that the human anima, together with the intellect, also contains the lower animas with their non-intellectual powers, thus constituting a functional "whole". Therefore, the same neurophysiological principle defined previously, i. e. the fusion of older with newer properties, must be considered as the principle operating at all levels of organization. Then, and here is the point, if we accept it for Man and the levels lower than the human, there is no reason for rejecting it when Man and levels higher than the human are involved, where the intellectual anima of man participates in some of the properties: "Est et alia ratio quare anima humana abundat diversitate potentiarum: videlicet quia est in confinio spiritualium et corporalium creaturarum, et ideo concurrunt in ipsa virtutes utrarumque creaturarum" (Sum.Theol. Ia, q.77, a.2) (see Note 22).

The same conclusions, in their essence, are reached by modern students of Thomas, interpreting in the light of present scientific knowledge the beautiful intuitions of the medieval philosopher. Modern theologians, however, still suspiciously regard the "continuum" (and the associated field theory) as a kind of undefined "sink" in which every individual must converge, every determination and definition is diluted to the extreme, to the loss of individuality, where there are no longer definite borders. However, the continuum does not reject the individual with its ensemble of properties; simply, the individual represents the anisotropy, i. e. the place of the heterogeneity of properties. In quantum physics, the sudden jump of the values of some variable of the continuum creates a singularity. Out of the continuum may arise the particle, the individual which moves throughout the continuum in an ordered relationship to it. In physics, the corpuscular and wave properties are not in opposition, but are complementary. Therefore, the concepts, the problems, and even the mathematical solutions, perhaps, of the physics of continua, of field theory, the whole science of today, would seem to adapt to the theological vision of the order of the Universe in Antiquity and in the Middle Ages. Surely, these arguments still allow a purely mechanistic explanation of the phenomena within the context of self-sufficient nature, only incidentally congruent with the older and unscientifically expressed thoughts, but, on the other hand, one may also surmise that science has a theological foundation.

### Notes

*Note 20.* Therefore, one must affirm that in Man there is no substantial principle besides the intellectual anima, and that, as the latter contains the nutritive and the sensitive animas, it contains in the same way all the inferior principles, and does alone everything the other less-perfected principles do in other beings.

*Note 21.* According to the order of Nature ..., the intellectual anima occupies the lowest level among the intellectual substances.

*Note 22.* There is another reason why the intellectual anima of man abounds in different powers, i. e., because it is at the borderline between the spiritual and the material creatures, and, therefore, the virtues of both concur in it.

## A. 5 Theology and Neuroscience

By proposing a question like the latter we are not proceeding far: it is obvious that it cannot get a satisfactory answer. Moreover, the point is not to prove one or another proposition. The purposes here are much more circumscribed. We have seen that Nature is self-consistent, in the words of one of the major students of Thomas, A.-D. Sertillanges: "Le succès de la méthode éxpérimentale appliquée aux vivants prouve, étant donné les moyens de cette méthode, que la vie a pour instruments éxclusifs de son action les forces physico-chimiques". Thus, until we probe living matter with the probes for physicochemical forces, we shall always reveal independent self-consistency in all natural manifestations. But, for our medieval theologist, and his commentators, there is another level, that of the principles. There is a "substantial principle" which makes from the elementary qualities a unity, a whole apt to follow its vital cycle, a "composite" having properties not contained in the elementary components (see the quotation "Forma mixti corporis ...", Note 13). Since the basic principles of the theory of continua seem fit to physically express the general scheme of the medieval theologist, we shall probe the brain with them, not aiming to further describe its machinery, but to ascertain whether a trace of that substantial principle could be identified, In other words, whether we could pry open the door to that other level to which neither physics alone nor chemistry pertain. Conceded that the physicochemical laws form a unity themselves at a lower level (there being many types of interaction and cooperation between the elements of matter), what can unify them at a higher level must interact in some way with the unity at a lower level and, therefore, a trace of that principle could be detected.

Let us start by considering the continuum of the brain (as we have seen it in Chaps. 9, 10 and 11) in which there are fields, and lines of force in the fields. This means that there are stresses and strains, deformations of the space in different directions continuously changing with time. The matrix of the stress and strain tensors, or more simply, of the metric tensor, should represent all the patterns of change. The first point is that a "tension" of the continuum, as represented in the tensor's components, is always present. Its neurological substrate is the continuous activity of every neuron (what we have called the "spontaneous" or "ongoing" activity), present, as we have seen, for the very reason that the neurons are interconnected. The functional property we have called "correlation" between the activities of groups of neurons, is the deformation of the phase space. However, the deformation of the continuum, induced by those forces which modify the correlation, itself changes the field of the forces and their lines, i.e. there is interaction, as we have previously described, between the stress and the strain tensors.

Granted that tension is always present, let us concern ourselves with its properties. A first observation is that the tension is present even if there are no disturbances from outside the brain, which by no means excludes external intervention in the maintenance of ongoing activity. Perfect isolation of the brain from external stimuli is an artificial situation, hardly achieved at all in specialized laboratories; moreover, it is utterly impossible to prevent the body stimuli from reaching the brain. Therefore, what we call spontaneous or ongoing activity always concerns the processing of information originated within, but also outside the brain. There is direct evidence that in the newborn, and before birth, the neurons are active, when the brain is dealing, almost exclusively, with information from inside the body, a kind of processing of which even the adult generally gains no conscious experience. This leads to the conclusion that the earliest activity, or tension in our brain (the distribution of the field intensity and the lines of force at birth and before), cannot be related to the highest, or intellectual, properties of the nervous organization. This was already an accepted truth in the Middle Ages: Thomas states that in the course of generation "... prius embryo habet animam quae est sensitiva tantum; qua abiecta, advenit perfectior anima, quae est simul sensitiva et intellectiva" (Sum.Theol. Ia, q.76, a.3, ad 3m) (see Note 23), because, he adds, "... imperfectiores potentiae sunt priores in via generationis: prius enim animal generatur quam homo" (Sum.Theol. Ia, q.77, a.7) (see Note 24); and, quoting Aristotle, concludes "Intellectus ... in principio est sicut tabula rasa in qua nihil est scriptum. Quod manifeste apparet ex hoc, quod in principio sumus intelligentes solum in potentia, postmodum autem efficimur intelligentes in actu" (Sum.Theol. Ia, q.79, a.2) (see Note 25). As for the latter quotation, of course, neither Aristotle nor Thomas could foresee the results of modern EEG research, so it is obvious that they were referring to an operation, that of the anima intellectiva, which only later begins to analyse the external world and judge external events. The full development of this function is certainly delayed in time, requiring maturation of the structure and its chemical properties, and, therefore must be completed somewhat later in the life of the individual.

The problem of maturation is not uniquely one of the brain, but of the body as well, including its sense organs. There are a number of places where Thomas clearly states a temporal priority of external information in evoking the intellectual operation of the brain: (quoting Democritus) "Nulla est causa cuiuslibet nostrae cogitationis, nisi cum ab his corporibus quae cogitamus, veniunt atque intrant imagines in animas nostras" (Sum.Theol. Ia, q.84, a.6) (see Note 26). The body with its sensors, therefore, introduces the images, but: (quoting Augustine) "... corpus non sentit, sed anima per corpus, quo velut nuntio utitur ad formandum in seipsa quod extrinsecus nuntiatur" (Sum.Theol. Ia, q.84, a.6) (see Note 27). We have, then, the first step in the process of formation of ideas, i.e. the abstraction: "... quando aliquis conatur aliquid intelligere, format aliqua phantasmata sibi per modum exemplorum, in quibus quasi inspiciat quod intelligere studet ... Et ideo necesse est ad hoc quod intellectus actu intelligat suum obiectum proprium, quod convertat se ad phantasmata, ut speculetur naturam universalem in particulari existentem (Sum.Theol. Ia, q.84, a.7) (see Note 28). Hence, the conclusion: "Secundum hoc, ergo, ex parte phantasmatum intellectualis operatio a sensu causatur. Sed quia phantasmata non sufficiunt immutare intellectum possibilem, sed oportet quod fiant intelligibilia actu per intellectum agentem; non potest dici quod sensibilis cognitio sit totalis et perfecta causa intellectualis cognitionis, sed magis quodammodo est materia causae" (Sum.Theol. Ia, q.84, a.6) (see Note 29).

The kernel of Thomas's idea is that up to some very early stage of development the brains of Man and of animals should perform on more or less the same level, being on an approximately comparable level of organization (corresponding to the powers of the sensitive anima, which in Man appear and mature earlier than those of the superior, intellectual anima). But, as maturation proceeds, the operations fostered by the incoming information have progressively less sensory content and a more intellectual one, until a pure intellectual operation "explodes". With this, Man (and his brain) can be said to be completed, as quoted earlier: "Virtus intellectiva non est virtus alicuius organi corporalis ... Sed in materia est, inquantum ipsa anima cuius est haec virtus, est corporis forma, et terminus generationis humanae" (Sum.Theol. l.c., Note 10). The point of distinction between animal and human proficiency should be the appearance of a specific capacity to extract the essence from the minuteness of data produced by analysis, as quoted earlier: "... tamen est quaedam operatio animae in homine quae dividendo et componendo format diversas rerum imagines, etiam quae non sunt a sensibus acceptae" (Sum.Theol. l.c., Note 8). This property should mark a noticeable difference in the tension distribution between the brain of Man and that of animals. Is this very different from what, not knowing the philosophers and theologians of the Middle Ages and antiquity, we would reasonably expect, following our science-based knowledge? I dare to say it is not. No ethologist would deny the abyss separating the performances of human and animal brains, nor that the difference resides in the intellectual power of man, or that the latter takes longer to fully develop, but that different behaviors are always to be expected.

However, the crucial point is still something else. Intellectual power could be something that does not slowly develop, but critically appears, as I have just said, "explodes". There are some arguments for this point. Every animal reaches some level of knowledge of itself, of its companions and mates, of the world around, through the information introduced by its sensors, through learning and practicing, a knowledge not classified as intellectual, being not autonomous, not "dividendo et componendo", limited to the immediate necessities and contingencies of its life, largely unconscious. In order to reach an intellectual knowledge as in Man, there is not a quantitative jump, nor there is evidence of a difference of level which might be overcome by degrees. Rather, there is a qualitative jump to perform, and there are no intermediate qualities. Thus, up to a given level of maturation, and assuming the same informational content in the sense organ messages, the tensors of the brain space in animals and Man could show comparable transformations induced by the incoming messages, but beyond some critical level the tensors should transform in a different way; there should be different tensors. The existence of such a qualitative jump was accepted by the medieval theologian, who foresaw an "agent" for it: as quoted previously "... Sed quia phantasmata non sufficiunt immutare intellectum possibilem, sed oportet quod fiant intelligibilia actu per intellectum agentem ...," (again see Note 29). What is actually meant by the first of these two intellects , is: "... intellectus qui est in potentia ad intelligibilia, quem Aristoteles ob hoc nominat intellectum possibilem" (Sum.Theol. Ia, q.79, a.2) (see Note 30). The second one, then, is what really identifies Man: "Oportebat igitur ponere aliquam virtutem ex parte intellectus, quae faceret intelligibilia in actu, per abstractionem specierum a conditionibus materialibus. Et haec est necessitas ponendi intellectum agentem". (Sum.Theol. Ia, q.79, a.3) (see Note 31). The concept is expanded and reinforced: "Intelligibile autem in actu non est aliquid existens in rerum naturam, quantum ad naturam rerum sensibilium, quae non subsistunt praeter materiam. Et ideo ad intelligendum non sufficeret immaterialitas intellectus possibilis, nisi adesset intellectus agens, qui faceret intelligibilia in actu per modum abstractionis." (Sum.Theol. Ia, q.79, a.3, ad 3m) (see Note 32).

The agent-intellect (intellectus agens), which essentially promotes abstractions, must be the principle originating that property peculiar to man – language, the property we have seen develop and express complex symbolisms of different natures, at the very basis of knowledge itself, the beginning of the route to Science and Truth. The question for Thomas was the same as for us: both, animal and man receive the same "impression" from the senses, but only man attains complete knowledge because something critical appears in him, "agent-intellect", which makes it possible for the impression to be assimilated (known) to the point that it becomes part of his being, of his intellect, by modifying it. This is why Aquinas (and Aristotle before him) said that the intellect, initially, is "sicut tabula rasa in qua nihil est scriptum" (l.c., Note 25). Nothing can be written on it, nothing can change it until the agent-intellect sets in action the powers of the potential intellect, only then can knowledge begin.

The comment by A.-D. Sertillanges is more clear than my words: "En ce sens, il est vrai de dire que la connaissance intellectuelle n'est que de la sensation transformée; seulement, la transformation est totale. On dirait de même que la sensation n'est qu'une alteration physico-chimique transformée; seulement la transformation est totale. Il y a là divers ordres. Pour passer de l'un à l'autre, il faut bien supposer un pouvoir. Ce sera ici l'activité du sens; ce sera là ce pouvoir de l'intelligence révélé et couvert un peu par ce nom: l'intellect agent".

Sertillanges uses in his comment the same term "transformation" I have used throughout the preceding chapters to indicate the changes induced in the continuum of our brain space by a tensorial operation. Although, at that time, he could not have had the necessary knowledge, he probably would have also accepted the modern concept of "operator" as the agent of transformations. Accordingly, the action of the agent-intellect would be manifested through an operator inducing in the matrix of the tensors a change different to that of the sensory operator. However, whereas the operator of sensation would be utterly dependent upon the content of the sensation, and also time-dependent in many aspects, the operator of intellectual knowledge would have none of these properties, representing something quite different. The power of intellectual knowledge is unique and irrevocably obtained. its appearance (compared to the duration of life) could be compared to a step function, its effects remaining stable. It would be hard to identify the time of its appearance, whereas its effects, being everlasting, should theoretically be revealed by some steady deformation of the fields of force and their lines. At what time in the life of Man this change appears, is not easily predictable. It might seem, since after all every increase of brain performance requires at least the maturation of structure, that it should occur after maturation is completed, i. e. after the first few months of life. However, this operator might well intervene much earlier, as one of the factors fostering maturation, thus actually appearing, or being already present at birth. Only its outward signs would appear later, i. e. the maturation of behavior, which requires the full development of the brain's structure and its relationships to the body. Certainly, all the morphological components of the difference between humans and animals are included in the genetic transmission of the characteristics of the species, but, considering the much wider variety of behavior in humans than in animals, the genetically defined, functional component cannot be overwhelming, thus letting each individual man enjoy a much greater degree of freedom than animals do. However, let us withhold our opinion until the next two points have been discussed.

## Notes

*Note 23*. ... Initially, the embryo has only a sensitive anima; once it has been shed, a more perfect anima appears, which is simultaneously sensitive and intellectual.

*Note 24*. ... The more imperfect powers are generated earlier: in fact, the animal is generated earlier than Man.

*Note 25*. The intellect ... initially, is like a white page upon which nothing is written. This is apparent in the fact that initially we are only potentially intelligent, later on we become intelligent in actual fact.

*Note 26*. There is no cause for our thought, other than the images which proceed from the material bodies about which we are thinking, and enter our animas.

*Note 27*. ... The body does not feel, but rather, the anima feels through the body, which it (the anima) uses like a messenger, in order to reconstruct in itself what is perceived from the outside.

*Note 28*. When someone tries to understand something, he constructs within himself, as examples, imaginary representations (phantasmata), by which he may, in a sense, examine what he is endeavouring to understand. Therefore, so that the intellect may actually understand its object, it is necessary that it revert to these imaginary representations, so that it may contemplate the universal essence existing in the particular.

*Note 29*. Accordingly, then, through imaginary representations, an intellectual operation is produced by the senses. However, since imaginary representations are not sufficient to introduce a new state in the potential intellect, it being necessary that they become actually intelligible by means of an agent-intellect, one cannot affirm that sensory cognition is the complete and perfect source of intellectual knowledge, but rather, that in some way it gives substance to the source.

(Translating "intellectus agens" into "agent-intellect" stresses the point that the latter is the intellect which "achieves", by transforming in actuality something which is there only in potentiality).

*Note 30*. The intellect which is potentially suited to understanding, the property Aristotle called the "potential" intellect.

*Note 31*. It was necessary, then, to assume some property of the intellect which would cause things to become actually intelligible by separating essential qualities from material things. This created the necessity for assuming an "agent-intellect".

*Note 32*. The property of being actually intelligible is not something inherent in the nature of things, as much as in the nature of perceptible things cannot exist outside matter. For this reason, the spirituality of the potential intellect would not be sufficient for intellectual cognition, were it not for the presence of the agent-intellect, which makes things actually intelligible through its capacity for abstraction.

## A. 6 Tension

The more we consider our brain with its innumerable components, all simultaneously active and tightly bound together while performing functions, the more we perceive that the continuum (of whatever properties or functions), constructed from fields of force with their lines or preferential paths, has in itself the reason for its own existence. We have seen that a tension of our brain's continuum is always present, even at birth and before, revealing in its lines the ongoing processing of information (which at a very early time is probably concerned only with bodily functions). However, at all ages, this is a tension impressed by events outside the brain on some substratum which is already "prepared" to receive it. That there must be a previous preparation, what it may consist of, and the reasons behind it might, alone, be taken as the cornerstone of an articulated theory of brain function. It is worthy, then, that we spend some time on it.

The root of this hypothesis will be more clear if we connect our arguments to what has been said in the preceding section about the operator of intellectual knowledge. In the process of transformation of sensation, by which the invariant nature of the latter, its essence, becomes part and parcel of our intellect, is also included the recognition of its "significance" for us, i. e. of the place the new knowledge is going to have in our future behavior. Only through newly acquired knowledge, is the complex of our acts, our behavior, consciously, purposefully and not simply mechanically modified (we might as well again use the term transformation for our behavioral changes). Here is the point: Man can recognize that significance only by comparing it with something in his possession. The question is, what standards does he possess and use for that comparison? There seems to be only one possible hypothesis, that the substrate of his brain has been somehow "pre-adapted", prepared to receive the content of the transformations, over the whole range of his powers. If we accept that in man's brain something of the brute's being, with all the inherent drives of animal life, coexist with the intellectual being, also with its inherent drives, then, what I called "preparation" should be an original, primitive, basic tension, upon which all the other tensions, resulting from whatever transformations, are later superimposed. So, the reason for the original tension, its purpose in a teleological sense – the finalistic view has its proper place in every theory – can be found only in that it may represent a term of comparison for all the possible "pulls" of life after their transformation. The fields of force should have, thus, a structure to start with, determining the basic strain in the continuum. Everything intervening later either follows in full or in part the pattern of the existing lines of force, thus reinforcing the basic deformation, or somehow contrasting with it. It is merely the procedure of comparison which, in this respect, is the basic process of the nervous system, the first to appear as intellectual power, through which the being starts "judging" (the comparison leads to judgement) the external world and his own internal reactions (the transformations of it) with his own measuring "rod". In this sense too, we can interpret the statement of Aristotle and Thomas, that the intellect at birth is like a "tabula rasa", a white page, that the human intellect knows nothing of the world around, that there is no inborn memory. What enters through the senses and is subjected to transformation is not compared with the content of a genetically printed memory, but is confronted with a prepared, "directed" intellectual power, which judges animal (brute) drives and intellectual (human) pulls and transformations with its peculiar, human comparison standards. It would be a uniquely human situation. Everything is left to his judgement: "Deus constituit hominem et reliquit eum in manu consilii sui" (Eccl. 15 ⟨14⟩) (see Note 33), a proposition acceptable outside any religious belief.

The situation would seem to be altogether different in animals. In them, the ongoing activity "must" be related to inborn behavior, to genetically printed patterns, or memories, which do not direct judgement, which the brutes do not have, but the action of the beings from the very beginning of their life, in order that the animals may perform correctly from the start, with only minor faults due to lack of practice. From this viewpoint, the basal tension would have a completely different significance, like a book in which the animal reads its own life, its past (as its predecessors have written it), which shall become its future, in an immutable succession of mechanically repeated behaviors, only changeable by genetical mutations, by slowly developing adaptations to environmental conditions, and by a few learned or imitative acts. In the basic tension of the animal brain's continuum there would be a fully determinative significance, an absence of freedom of the individual and of the species, making the animal an integrated element of the immutable and predictable part of nature. It is not to be wondered at that the neurophysiologist, investigating the animal brain as a "model" of the human one, has gained the impression of an unavoidable, mechanical development of the nervous processes, which he uses to logically justify his empirical, mechanistic research philosophy. Also, when he transfers his investigations from the brain of a rat to that of a monkey, enormously more complex and organized, the situation is not substantially changed, the primate's behavior still being sufficiently predictable, in spite of the increased number of intervening variables.

It is with Man that the neuroscientist faces unsurmountable hindrances to the "good-for-everything" empirical, mechanistic approach (for him, the only approach) and has to base the principles of modern experimental psychology on statistical calculations alone. This is what led Karl Jaspers to deny the possibility of building a general theory of psychopathology, but clearly, that judgement could be generated only by an unsuitable approach. The empirical, mechanistic approach is the typical philosophy of analytical investigation, which explains the mechanisms of and the reasons for phenomena on the basis of the fact that for each there is one antecedent, which (by virtue of structural connections, and physical and chemical properties) determines the following ones. This is the approach which makes the evolution of events, and their machinery understandable, but it is unfit for unveiling those causes which reside in the primary logic of the living being, in the principles of organization, and the laws on which the physical corpus of a theory is built. As Einstein said, there is no logic way which may lead us to build a theory starting from experience. If we want to understand Man, for whom the tension of his continuum is the original mark of his intellectual powers, of his ability to judge by his own measuring standards the transformations of the sense organs' information, for whom, therefore, there are not uniquely defined antecedents, necessarily determining their consequences, we must abandon the empirical, mechanistic approach and start the other way around, with the principles, i. e. with a theory. In the latter proposition there is much more than a recipe for a sensible approach to research: if accepted, there is the aknowledgement of the role of primary logic in determining the principles of organization, which in turn are expressed in the laws of the physical structure embodying them. From this, we can draw only one consequence, i. e. that the structure and the functional properties with which the structure is endowed are determined by that primary logic. Once more, we find that an identical concept is contained in the medieval philosophy, clearly stated in Thomas's words: "Non enim potentiae sunt propter organa; sed organa propter potentias: unde non propter hoc sunt diversae potentiae, quia sunt diversa organa; sed natura instituit diversitatem in organis, ut congruerent diversitati potentiarum" (Sum.Theol. Ia, q.78, a.3) (see Note 34). Which means that the original logic, the principle (the anima with its powers) is the one determining both the structure and its organization, and thus the function of the resulting organ, the brain. In the human, what distinguishes him from animals and makes him a man is the intellect, which expresses itself in abstractions, symbolization, thought, rationality of judgement, etc., powers of that "something" the philosopher calls "anima"; and for the philosopher the brain is the way it is in order to be congruent with those powers, i. e. it can only have been determined by them. For the philosopher too, in order to understand Man one has to begin with his essential feature, his intellect, because there resides Man, and there are the causes of his material life as well.

## Notes

*Note 33.* God created Man, and left him in the power of his own judgement.

*Note 34.* In fact, the powers do not exist on account of the organs. Rather, the organs exist because there are powers: hence, different powers do not exist because there are different organs; but Nature created differences in the organs in order that they might be congruent with the differences of the powers.

# A. 7 Determination and Free Will

Talking of determination we come to the last point I want to make in this Appendix. As we have seen at the beginning of the book, Man and animals behave differently in the same situations. In everyday life Man and animals perform activities common to practically all living beings, such as, for instance, perception of the external world, recognition of familiar objects, satisfaction of hunger and thirst, etc. Everyone would be ready to concede that these are simple acts of life performed similarly by animals and humans. Nonetheless, the actual similarity is restricted to a very few examples and is much more superficial than previously thought. Man actually puts into his deeds a "quality" all his own, in all the situations of life. In no occasion does Man do quite the same things as animals do. He performs the same functions (feeding, mating, sensing, etc.) but not in the same way. We might say that he does them in a "better" way, but essentially, he does "different" things. There is a qualitative difference, not a quantitative one. As we have seen, the reason is to be sought in the fusion of primitive structures and organizations with newer and more evolved ones, so that very simple acts, like feeding and mating, can no longer be accomplished by Man in the same way as by animals. Those intellectual powers which distinguish Man also appear in his simplest and animal functions, as different ways of conceiving and executing them. For example, Man perceives feeding and mating as part of a life which includes, surely, also these functions, however, only as part of an extremely wide complex of human functions, in which they are deprived of some of that compulsive significance typical of animal life, while they are endowed, at the same time, with a more intellectual (or spiritual) human connotation. The concept has been emphasized by one of the modern students of Thomas's thought, M.-D. Chenu, whom I should like to quote: "Certes, l'appétit sensitif comme tel est régi par l'instinct; mais, dans le composé humain, il participe en quelque sorte à la dignité de la raison, – mirabilis connectio – au point que les mouvements non déliberés de cet appétit tombent sous la moralité, dès là que en principe ils sont sous l'empire de la raison." I have introduced this quotation because it points out the philosophical side of the fusion of the sensitive with the intellectual powers in the brain's organization (the "mirabilis connectio"), but, principally, because it raises problems which are fundamental in our conception of Man, of his standing in the world, that is, the problem of "freedom", the other side of the question of determination and responsibility, the moral problem. We have encountered "free will" at the end of Sect. A. 2, and we are going to dwell on it for a while.

The predictability of inanimate Nature as to its physicochemical laws which govern all matter, can be extended to a considerable degree, though certainly expressed in much more complex relationships, to the living realm. We have recognized this property by saying that Nature is self-consistent, that the mechanistic development of the physicochemical laws practically accounts for all the phenomena related to the life of the organs, even those developing in the more evolved brains of subhuman primates. Predictability is merely the outcome of determination, the fact that an antecedent is always followed by a known consequential chain of events, and once the latter are known, so also are known the conditions of the antecedent.

I have briefly commented on chance and determination in Sect. 6.1.1. It was actually one of the main reasons for writing this Appendix. Let us now put the problem in these terms: granted that animals are easily predictable, they are so because of the anatomical structure of their brain which only allows that degree of organization (in metaphysical terms, only a sensitive anima is allowed). Determination, therefore, is a problem of structure and its organization. As a consequence, we ought to conclude that, since Man's brain also has its own (more complex) structure, allowing a higher level of organization, in his superior complexity of structure and organization Man should also be determined and, therefore, predictable. That some intellectual power of the human brain is subject to determination has been assumed in the preceding sections, being described as an original basic tension, representing the intellectual standard of comparison, the one that Man has no possibility of building by himself, being his inheritance at birth, part and parcel of his intelligence, of his anima intellectiva, of his specific nature. That Man, too, may exhibit predictability at some level, as animals do, is also true, though to a limited extent, even when the predictability of Man is probed with functions typical of the "anima sensitiva", for instance, thirst satisfaction. In this situation, which corresponds to a basic need of life, we would easily predict that a thirsty man will drink, since his brain is responding to the outside stimulus and his behavior can reasonably be expected to depend upon it, exactly like that of a cat. But there are also many situations in which a thirsty man does not touch the water in front of him, even if strongly urged by thirst: in the words of Thomas, previously quoted, "Sed homo non statim movetur secundum appetitum irascibilis et concupiscibilis, sed expectatur imperium voluntatis, quod est appetitus superior" (Sum.Theol. l.c., Note 9), and again "Necessitas autem naturalis non aufert libertatem voluntatis" (Sum.Theol. Ia, q.82, a.1, ad 1m) (see Note 35). As a matter of fact, no one can deny that Man can actually behave quite unpredictably, even if the prediction is based on the presence of strong drives, such as the longing for satisfaction of a basic, natural need. As I said, closing Sect. 3.4, a voluntary act can interrupt a deterministic chain of events, such as that set in motion by thirst. However, we can also say that the very refusal of water, unpredictable as it may appear, must have actually been determined by some particular situation, e. g. when there is not enough water to satisfy all the people present, which can induce a man to stop drinking. So, it would again be, after all, a dependence upon factors which, once known, can render the behavior of Man predictable, thus revealing an actual determination. Since there is nothing in the sensitive anima capable of interrupting the deterministic chain of events ("ovis timens lupum statim fugit" says Thomas, l.c., Note 9), the possibility of a different choice, actually of "a" choice at all, when particular situations arise, can reside only in the superior complexion of structure and organization which embodies the powers of the intellectual anima, a peculiarity of Man. In other words, Man's peculiar behavior is made apparent by his independence from external drives, or by his ability to substitute some internal drive of his own for the external drive present.

Let us consider, first, what happens when Man isolates himself from his surroundings, from all that may exert a pull from without, i. e. (to continue with the image of the rubber sheet) from what represents the stresses for the brain continuum, the environment's messages to his brain. No one wonders whether a man in relative isolation succeeds, for instance, in developing a new philosophical system or other products of his intellect, because if it is true that, as I quoted earlier "... non potest dici quod sensibilis cognitio sit totalis et perfecta causa intellectualis cognitionis, sed magis quodammodo est materia causae" (Sum.Theol. l.c., Note 29), then Man has other sources of matter for his thought, and "dividendo et componendo" "multa cognoscit quae sensus percipere non potest" (Sum.Theol. l.c., Note 7). Man is capable of abstracting the essentials from many objects and symbols, constructs ideas, makes music from sounds, poems from words and paintings from colours, has created languages, mathematics and philosophies, investigates mechanisms and causes in Nature and in himself, and steadily builds Science. Man's brain seems to perform better, i. e. best develops its intellectual functions, when it is not "disturbed", when its "rubber sheet" is least stretched by those events which seem to be the source of drives in animals. No animals show clear-cut signs of independent, autonomous thinking, or of creating anything; they only seem to live in the world, and they live and multiply for the world, being part of the equilibria of Nature. The tension in their brain, thus, is maintained by the world of objects, by their worldly affairs, food, companions, reproduction, preys and predators. When isolated from the world they can simply live, their brain in that condition should not be preferentially pulled in any definite direction, being only kept ready to respond to the environment's stimuli by the ongoing activity of their neuronal circuits, only concerned with bodily information, a basic tension, but of a lower grade, an engine at its minimum speed, as we have previously defined.

From the neurophysiological point of view, the stress and strain tensors of the continuum in animal brains should exhibit a fixed distribution, defined mostly by heredity and least by experience, which must await external stimuli in order to change and start to organise corresponding behavior. Metaphysically, this limitation, this strict dependence upon the external world, is the expression of the anima sensitiva, of the animal essence, which disappears with the death of the body. However, with humans things seem to be different, a situation already present at the dawn of Mankind, when Man knew nothing of the world, apart from a handful of his fellows and some huge beasts, when he had no scientific knowledge or philosophies at his disposal to help him with a hostile Nature. Together with all the world's strains he must have also experienced other strains, which made him strive towards the "outside-the-surroundings", beyond the visible and the touchable. However, this could not be taught to him by the world, there being nothing in the matter around him which could suggest "no matter". So, the stresses and strains in the "rubber sheet" of his neuronal circuits, the tension which urged him to build his knowledge, must have been the expression of another type of essence, present from the very beginning, his "anima intellectiva", his power to attain knowledge about what his senses had no cognition of, to judge about an "innermost" of himself and a world that could not be seized materially. This makes the human a living being in which the basic tension has definite (or preferential?) directions, on one hand towards animal life, the animal nature, defined by all the strains of the world inducing corresponding strains in his brain, and, on the other towards the "beyond-the-world", the world of pure knowledge, of spiritual life, a domain constituting an everyday concern for Man, but not for animals. It is the latter domain which has created Science, although the natural scientists of today are unwilling to let themselves be involved in its problems, escaping as they do the physicochemical probes, notwithstanding their continuous interference with Man's behavior. The positivistic science of the last two centuries has ended up considering these problems as being not "natural", but useless or superfluous, artificial constructions, in some way added to complicate Man's life with strange ideas about good, evil, moral, religion, philosophies, etc.

Surprisingly, these considerations reinforce the belief in a deterministic principle rigidly governing the acts of Man as those of animals, certainly at different levels, with different means and in different directions, but a determination all the same. In fact, if the brain's structure is the way it is in order to be congruent with the powers, then there must be a source of determination for the powers as well. Here is the real problem: that of the freedom of Man, of his "liberum arbitrium", of his free will when we see him interrupting the course of the mechanistic development of the events in which he is involved, and giving them another course, not foreseen by the material cause and effect relationships.

In point of fact, the problem is not one of determination or free will, since determination can be opposed only to chance. We must realize that we are always "determined" in our actions, both when we let the mechanistic chain of cause and effects guide our actions, as well as when one or another of those strange principles (ethics, religion, philosophy, etc.) is our master: the free will, actually, consists only in choosing which of the ways shall provide guidance. This is the conclusion of the medieval philosopher too, albeit reached in other ways: "... sumus domini nostrorum actuum secundum quod possumus hoc vel illud eligere. Electio autem non est de fine, sed de his quae sunt ad finem." (Sum.Theol. Ia, q.82, a.1, ad 3m) (see Note 36). In fact, the common goal of all the different ways is the "good" (for the subject) which, however, can be seen in opposite ways. Our will is always determined toward what we believe (judge) to be "our" good. This is also clearly stated by Thomas: "... voluntas in nihil potest tendere nisi sub ratione boni. Sed quia bonum est multiplex, propter hoc non ex necessitate determinatur ad unum." (Sum.Theol. Ia, q.82, a.2, ad 1m) (see Note 37).

Thus, we have choice, the possibility of selection, but if we want the choice to be really free, it ought not to be immediately predictable, not exactly defined by the terms of the dilemma we are confronted with. The sentence previously quoted "necessitas autem naturalis non aufert libertatem voluntatis" (l.c., Note 35) states that a voluntary act can oppose a natural drive in Man's behavior, so that Man can, if he wants to, follow one of the many other drives which may urge him to action. However, in the case of non-natural drives as well, even particularly attractive ones, such as those which result from rational arguments anticipating immediate and appealing advantages, the choice is not uniquely determined, so that the behavior may even appear to be "irrational". Again Thomas: "... appetitus sensitivus, etsi obediat rationi, tamen potest in aliquo repugnare, concupiscendo contra illud quod ratio dictat." (Sum.Theol. Ia, q.83, a.1, ad 1m) (see Note 38). Nonetheless, also the decision whether to follow the impulse of a natural drive or not is a conscious and rational choice, because we always tend to do what we judge is the best for us, even if, sometimes, only for an immediate and transient situation. Therefore, in Man choice is always dependent upon an act of judgement. The conclusion is that free will and free judgement are synonyms, characteristic properties of Man alone, judgement being an act of the intellect, of his rational powers. The conclusion is beautifully stated by Thomas: "Sed homo agit iudicio: quia per vim cognoscitivam iudicat aliquid esse fugiendum vel prosequendum. Sed quod iudicium istud non est ex naturali instinctu in particulari operabili, sed ex collatione quadam rationis; ideo agit libero iudicio, potens in diversa ferri. Ratio enim circa contingentia habet viam ad opposita ... Particularia autem operabilia sunt quaedam contingentia: et ideo circa ea iudicium rationis ad diversa se habet, et non est determinatum ad unum. Et pro tanto necesse est quod homo sit liberi arbitrii, ex hoc ipso quod rationalis est. "(Sum.Theol. Ia, q.83, a.1) (see Note 39). Again: "Voluntas est in ratione". (Sum.Theol. IIIa, q.18, a.2, ad 1m) (see Note 40). For the medieval philosopher and theologian, Man, through judgement, can judge his own place in Nature, and critically weigh his own intellectual complexion, and the value of the goal to be reached, together with the consciousness of his own destiny, of wordly and superior goods and evils: all this belongs to the "rationale" of his choice, so that the latter is never the outcome of chance.

Finally, we come to the last step: free judgement becoming free will manifested in action. The latter, however, does not immediately or necessarily follow judgement. Only when judgement and desire (appetite) coincide, can action follow undisturbed. Psychiatrists teach that mental conflicts may arise, severely endangering psychic life, whenever judgement and appetite cannot be reconciled for consensual, unitary action; it is actually an old truth: "Iudicium de hoc particulari operabili ut nunc, numquam potest esse contrarium appetitui." (De Veritate, q.24, a.2) (see Note 41). This, of course, does not imply a constraint as to the goal to be reached. The question is never about the goal, but about the means to reach it, as previously stated. The goal of the appetite is always unique, the good of the self, but there can be uncertainties in Man's mind about the nature of good, since it may have different aspects. On this basis, appetite and rationality may conflict. The "appetitus sensitivus" (the sensitive appetite) can refuse the rational judgement, thus letting Man exert his right of independent and autonomous choice, but, when this appetite complies with rationality, there is no possibility of error, and in this case one is sure to attain the good. Thomas, and Augustine before him, arrived at this conclusion long ago: "... hoc ergo est bonum quod homo non facit cum vult, scilicet non concupiscere contra rationem". (Sum.Theol. Ia, q.83, a.1, ad 1m) (see Note 42).

As neuroscientists we could stop here. At the beginning we have indicated the intellectual powers of his brain (his rationality first of all) to be the distinguishing feature of Man, and now we end with the statement that all his appetites should yield to rationality. The latter, therefore, is for the scientist the ultimate and highest term of comparison, the one which is always in the right. However, just as physiologists are never content with the results of an investigation, we must ask ourselves whether there is really nothing more (see Sect. 1.3). We should like, for instance, to have a test assuring us that our rationality does not go wrong when it judges our "true" good, or alternately, we should like to have some kind of a "paradigm" of that good, so that we may recognize its essence even if it appears disguised, and may compare it with the verdict of rationality. We have no objective evidence of such a paradigm within the brain, but in behavior (which is the result of the whole brain's action) we have plenty of evidence that Man, having weighed the odds, having consulted appetite and rationality, yet discarded both, can decide upon a third way, following neither irrational sensitive appetite, nor a logical, rational and consequential chain of thoughts, but the image of another good, not from the senses or from reason. However, if Man can have "images" of a good which is not from the material, sensory domain, nor from his rational powers, then we must again think of something built into his brain, or imprinted at some earlier time of life, perhaps, if not that paradigm of an "essential good", or the "essence of good", at least a hint of a direction to it. Its characteristic should be (if we want to account for the strange behavior of Man) that once Man has gained awareness of it, it gives him another choice. Thus, recognizing a supreme term of comparison in that "something" is also an act of the intellectual and rational powers of Man, but yielding to it or not might no longer be a question only of free will, but a choice pertaining to another category, the one we call (natural?) ethics: it belongs to the moral domain. (See the quotation of M.-D. Chenu: "... les mouvements non deliberés de cet appétit tombent sous la moralité, dès là que en principe ils sont sous l'empire de la raison"). Of course, again we can refuse to act accordingly, but we can do so only by contradicting (inborn?) moral order. However, there might be still more in it. I am not sure whether I am interpreting Thomas correctly, but a few quotations seem particularly fit for a last comment: "Sicut igitur formae altiori modo existunt in habentibus cognitionem supra modum formarum naturalium, ita oportet quod in eis sit inclinatio supra modum inclinationis naturalis, quae dicitur appetitus naturalis." (Sum.Theol. Ia, q.80, a1) (see Note 43). Further: "Ex eo igitur, quod homo est aliqualis qualitate naturali quae attenditur secundum intellectivam partem, naturaliter homo appetit ultimum finem, scilicet beatitudinem. Qui quidem appetitus naturalis est, et non subjacet libero arbitrio ..."(Sum.Theol. Ia, q.83, a.1, ad 5m) (see Note 44). The last quotation refers to an appetite that cannot be suffocated voluntarily, since Thomas says it is in Man as the anima with its powers, his intellect with its properties, however, perhaps, with an even more basic character. The philosopher leaves no doubts about what is meant by beatitude as the ultimate goal of this appetite: "Sunt autem quaedam habentia necessariam connexionem ad beatitudinem, quibus scilicet homo Deo inhaeret, in quo solo vera beatitudo consistit." (Sum.Theol. Ia, q.82, a.3) (see Note 45). If so, Man's choice would be, ultimately, a choice between God and himself, no longer a simple question of ethics.

I do not know how many of my fellow neuroscientists would consider the arguments of this Appendix as a reasonable basis for designing experiments, but I have tried to present the matter in a way consistent with the preceding chapters, i. e. as far as possible as a problem of comparisons. As a matter of fact, it has been suggested, previously, that the primary operation of the brain is that of comparing the transformation of new entering events with traces or patterns in the structure of the brain continuum, built-in, or imprinted early, or learnt. Congruence of patterns may mean effortless conclusions, quick decisions, uncontested actions; lack of congruence may mean, instead, conflict, procrastination, indecision, delayed (or no) action. It would all rest on our ability of neuroscientists to detect patterns within the vast continuum of the brain functions; to read the particular distribution of stresses and strains in the circuits of the continuum and in the matrix of the tensors; to bring to light the preferential directions (if any) of the neural space deformations and the effects of the outer strains; to unveil, finally, the human standard of judgement, the detailed processes of Man's choices. Whether all this will be possible cannot be predicted, but the challenge is worthy of all our endeavours.

## Notes

*Note 35.* A natural necessity does not suppress the freedom of will.

*Note 36.* ... We are masters of our acts, inasmuch as we can select either this or that. The choice, however, is not of the goal, but of the means to reach it.

*Note 37.* The will can move only towards what is judged to be the "good". But, since the good is manifold, for this reason it is not necessarily determined in a unique way.

*Note 38.* The sensitive appetite, even if submitted to rationality, can however rebel in some matter, strongly desiring the opposite to what rationality dictates.

*Note 39.* Man acts on the basis of judgement, because through knowledge he judges what is to be avoided or pursued. However, this judgement does not come from a natural instinct for operating in the particular, but from a rational conclusion. For this reason, Man acts on the basis of free judgement, because potentially he can be carried in different directions. In fact, reasoning about contingent matters may lead to different ways ... however, the particular acts to be undertaken are contingent: therefore, rational judgement about them can be differentially oriented, and is not uniquely determined. Consequently, it is necessary that Man is endowed with free will for the very reason that he is a rational being.

*Note 40.* The will resides in rationality.

*Note 41.* The judgement of what must presently be done in a particular case, can never be contrary to desire (appetite).

*Note 42.* Therefore, this is good, that Man does not act as he wants, i. e. that he does not desire what is against rationality.

*Note 43.* As in those endowed with knowledge there exist constructive principles higher than the principles of physical Nature, it is necessary that there is in them a disposition higher than the natural disposition, which we call natural appetite.

*Note 44.* For the same reason that Man is what he is because of some natural quality, which is cared for by his intellectual side, so naturally Man tends eagerly toward the final goal, i. e. beatitude, which is a natural appetite and is not subjected to free will.

*Note 45.* There are also matters having necessary connections to beatitude, i. e. those through which Man is intrinsically bound to God, in whom alone is the true beatitude.

# References

This is a selection of books from which the neuroscientist may obtain fundamental concepts and ideas. The references under headings 1 and 4 will be very useful to physicists for getting acquainted with the brain's problems. Few references from scientific journals have been cited, the list being restricted to those items necessary for understanding old and new data.

### 1. Neurophysiology and Control Systems


Adrian ED (1947) The physical background of perception. Clarendon, Oxford

Adrian ED, Bremer F, Jasper HH (eds) (1954) Brain mechanisms and consciousness. Blackwell, Oxford

Arduini A (1983) Continuous correlation between brain activities. Arch Ital Biol 121:187–203

Bleuler E (1960) Lehrbuch der Psychiatrie. Springer, Berlin Göttingen Heidelberg

Blinkov SM, Glazer II (1968) The human brain in figures and tables. Plenum, New York

Borbely A, Valatax JL (eds) (1984) Sleep mechanisms. Springer, Berlin Heidelberg New York Tokyo

Borko H (ed) (1962) Computer applications in the behavioral sciences. Prentice-Hall, Englewood Cliffs

Callaway E, Tueting P, Koslov SH (eds) (1978) Event-related brain potentials in man. Academic, New York

Communication Biophysics Group of Research Laboratory of Electronics, Siebert WM (1959) Processing neuroelectric data. MIT Press, Cambridge

Culbertson JT (1963) The minds of robots. University of Illinois Press, Urbana

Dixon WJ, Nicholson WL (eds) (1974) Exploring data analysis. University of California Press, Berkeley

Elsasser WM (1958) The physical foundations of biology. Pergamon, London

Fraisse P (1984) Perception and estimation of time. Ann Rev Physiol 35:1–36

Glaser EM, Ruchkin DS (1976) Principles of neurobiological signal analysis. Academic, New York

Hebb DO (1962) A textbook of psychology. Saunders, Philadelphia

Hobson JA, Brazier MAB (eds) (1980) The reticular formation revisited. Raven, New York

Jaspers K (1959) Allgemeine Psychopathologie. Springer, Berlin Göttingen Heidelberg

John ER (1967) Mechanisms of memory. Academic, New York

Jung R (ed) (1973) Central processing of visual information. B: Visual centers in the brain. Springer, Berlin Heidelberg New York (Handbook of sensory physiology, vol 7/3)

Livanov MN, Rusinov VS (1969) Mathematical analysis of the electrical activity of the brain. Harvard University Press, Cambridge

Rashewsky N (1960) Mathematical biophysics. Physico-mathematical foundations of biology. Dover, New York

Richelle M, Lejeune H (1980) Time in animal behaviour. Pergamon, Oxford

Rosenblatt F (1962) Principles of neurodynamics. Spartan Books, Washington D.C.

Schmitt FO, Worden FG (eds) (1974) The neurosciences. Third study program. MIT Press, Cambridge

Siebert WM (1959) *See* Communication Biophysics Group of Research Laboratory of Electronics

Smythies JR (1970) Brain mechanisms and behaviour. Academic, New York

Stevens SS (1951) Handbook of experimental psychology. Wiley, New York

Thompson RF, Patterson MM (eds) (1973) Bioelectrical recording techniques, part A. Academic, New York

Von Neuman J (1959) The computer and the brain. Yale University Press, New Haven

Wiener N (1961) Cybernetics. MIT Press, Cambridge; Wiley, New York

Yovits MC, Cameron S (eds) (1960) Self-organizing systems. Pergamon, New York

## 2. Physics, Thermodynamics, Information Theory, and Related Subjects

Blackman RB, Tukey JW (1958) The measurement of power spectra. Dover, New York
Born M (1969) Physics in my generation. Springer, New York
Brillouin L (1956) Science and information theory. Academic, New York
Carnap R (1958) Introduction to symbolic logic and its applications. Dover, New York
Denbigh KG (1958) The thermodynamics of the steady state. Methuen, London
Dirac PAM (1981) Principles of quantum mechanics. Clarendon, Oxford
Dole M (1954) Introduction to statistical thermodynamics. Prentice-Hall, New York
Fu KS (1980) Digital pattern recognition. Springer, Berlin Heidelberg New York
Katchalsky A, Curran PF (1981) Nonequilibrium thermodynamics in biophysics. Harvard University Press, Cambridge
Khinchin AI (1949) Mathematical foundations of statistical mechanics. Dover, New York
Khinchin AI (1957) Mathematical foundations of information theory. Dover, New York
Lichnerowicz A (1967) Eléments de calcul tensoriel. Armand Colin, Paris
Lindsay RB, Margenau H (1957) Foundations of physics. Dover, New York
Marois M (ed) (1969) Theoretical physics and biology. North-Holland, Amsterdam
Prigogine I (1967) Introduction to thermodynamics of irreversible processes. Wiley, New York
Shannon CE, McCarthy J (eds) (1956) Automata studies. Princeton University Press, Princeton
Simmonds JG (1982) A brief on tensor analysis. Springer, New York Heidelberg Berlin
Solodovnikov VV (1960) Introduction to statistical dynamics of automatic control systems. Dover, New York
Takàcs L (1960) Stochastic processes. Methuen, London
Thomas R (ed) (1979) Kinetic logic. A Boolean approach to the analysis of complex regulatory systems. Springer, Berlin Heidelberg New York
Wax N (ed) (1954) Selected papers on noise and stochastic processes. Dover, New York
Wichmann EH (1971) Quantum physics. Berkeley physics course, vol 4. McGraw-Hill, New York
Woodward PM (1957) Probability and information theory with applications to radar. Pergamon, London

## 3. Relativity and Related Subjects

Bondi H (1979) La relatività e il senso comune. Zanichelli, Bologna (translation of: Relativity and common sense. Anchor Books, Garden City, 1964)
Born M (1964) Die Relativitätstheorie Einsteins. Springer, Berlin Göttingen Heidelberg
Einstein A (1916) Die Grundlage der allgemeinen Relativitätstheorie. Ann Phys IV F 49:769–822
Einstein A (1967) Relatività. Esposizione divulgativa. Boringhieri, Torino (Translation of: Über die spezielle und allgemeine Relativitätstheorie, gemeinverständlich. 1916)
Gal-Or B (1981) Cosmology, physics and philosophy. Springer, New York
Heidmann J (1980) Relativistic cosmology. Springer, Berlin Heidelberg New York
Schmutzer E (1981) Relativitätstheorie – aktuell. Harri Deutsch Verlag, Frankfurt
Sciama DW (1979) La relatività generale. Zanichelli, Bologna (Translation of: The physical foundations of general relativity. Anchor Books, Garden City, 1969)
Straumann N (1981) Allgemeine Relativitätstheorie und relativistische Astrophysik. Springer, Berlin Heidelberg New York

## 4. Natural Philosophy

D'Espagnat B (1983) Alla ricerca del reale. Boringhieri, Torino (Translation of: À la recherche du reel. Le regard d'un physicien. Bordas, Paris, 1981; English edition: In search of reality. Springer, Berlin Heidelberg New York, 1983)
Husserl E (1965) Idee per una fenomenologia pura e per una filosofia fenomenologica. Einaudi, Torino (Translation of: Ideen zu einer reinen Phänomenologie und phänomenologischen Philosophie. Nijhoff, Den Haag, 1950–1952)
Nagel E (1981) La struttura della scienza. Feltrinelli, Milano (Translation of: The structure of science. Harcourt, Brace, New York 1961)
Popper KR, Eccles JC (1977) The self and its brain. Springer, Berlin Heidelberg New York

## 5. Appendix

Chenu M (1963) L'Homme dans le monde. In: Asssociation des Professeurs de Philosophie des Facultés catholiques de France (eds) Recherches de philosophie VI. Saint Thomas d'Aquin aujourd'hui. Desclée de Brouwer, Paris, pp. 171–175
Dubarle D (1963) Cosmologie thomiste et philosophie naturelle contemporaine. In: Association des Professeurs de Philosophie des Facultés catholiques de France (eds) Recherches de philosophie VI. Saint Thomas d'Aquin aujourd'hui. Desclée de Brouwer, Paris, pp. 137–169
Bergomo Petri de (1960) In Opera Sancti Thomae Aquinatis index. Alba, Roma
Ruffini F (1943) Studi sul Giansenismo. La Nuova Italia Ed, Firenze
Sancti Thomae de Aquino (1962) Summa Theologiae. Alba, Roma
Sertillanges A-D (1925) S. Thomas d'Aquin. Librairie Felix Alcan, Paris

MIX
Papier aus verantwortungsvollen Quellen
Paper from responsible sources
FSC® C105338

If you have any concerns about our products,
you can contact us on
**ProductSafety@springernature.com**

In case Publisher is established outside the EU,
the EU authorized representative is:
**Springer Nature Customer Service Center GmbH**
**Europaplatz 3, 69115 Heidelberg, Germany**

Printed by Libri Plureos GmbH
in Hamburg, Germany